𝔖it nomen 𝔇omini benedictum

The
Eureka Stockade

The Consequence of Some Pirates Wanting on Quarter-Deck a Rebellion

by

Carboni Raffaello

By the grace of spy "Goodenough," Captain of foreign anarchists
But by the unanimous choice of his fellow-miners
Member of the Local Court Ballaarat.

SYDNEY UNIVERSITY PRESS

SYDNEY UNIVERSITY PRESS

SETIS at the University of Sydney Library
University of Sydney
www.sup.usyd.edu.au

First published in Melbourne and printed for the author by J. P. Atkinson and Co. in
1855.

The publication of this book is part of the University of Sydney Library's Australian
Studies electronic texts initiative. Further details are available at
www.sup.usyd.edu.au/oztexts/

© 2004 Sydney University Press

ISBN 1 920897 40 2

For current information see http://purl.library.usyd.edu.au/sup/1920897402

NOTA BENE

IN person I solicit no subscription — in writing I hereby ask no favour from my reader. A book must stand or fall by the truth contained in it.

What I wish to note is this: I was taught the English language by the Very Reverend W. Vincent Eyre, Vice Rector of the English College, Rome. It has cost me immense pains to rear my English up to the mark; but I could never master the language to perfection. Hence, now and then, probably to the annoyance of my Readers, I could not help the foreign idiom. Of course, a proper edition, in Italian, will be published in Turin.

I have nothing further to say.

CARBONI RAFFAELLO.

Prince Albert Hotel, Bakery-Hill, Ballaarat.
Anniversary of the Burning of Bentley's Eureka Hotel, 1855.

———————

The Eureka Stockade.

I.

FAVETE LINGUIS.

Mendacium sibi, sicut turbinis, viam augustam in urbe et orbe terrarum aperuit.
Stultus dicit in corde suo, "non est Deus."
Veritas vero lente passu passu sicut puer, tandem aliquando janunculat ad lucem.

Tunc justus ut palma florescit.[1]

I UNDERTAKE to do what an honest man should do, let it thunder or rain. He who buys this book to lull himself to sleep had better spend his money in grog. He who reads this book to smoke a pipe over it, let him provide himself with plenty of tobacco — he will have to blow hard. A lover of truth — that's the man I want — and he will have in this book the truth, and nothing but the truth.

Facts, from the "stubborn-things" store, are here retailed and related — contradiction is challenged from friend or foe. The observation on, and induction from the facts, are here stamped with sincerity: I ask for no other credit. I may be mistaken: I will not acknowledge the mistake unless the contrary be proved.

When two boys are see-sawing on a plank, balanced on its centre, whilst the world around them is "up" with the one it is "down" with the other. The centre, however, is stationary. I was in the centre. I was an actor, and therefore an eye-witness. The events I relate, I did see them pass before me. The persons I speak of, I know them face to face. The words I quote, I did hear them with my own ears. Others may know more or less than I; I mean to tell all that I know, and nothing more.

Two reasons counsel me to undertake the task of publishing this work; but a third reason is at the bottom of it, as the potent lever; and they are —

1st. An honourable ambition urging me to have my name remembered among the illustrious of Rome. I have, on reaching the fortieth year of my age, to publish a work at which I have been plodding the past eighteen years. An ocean of grief would overwhelm me if then I had to vindicate my character: how, under the hospitality of the British flag, I was put in the felon's dock of a British Supreme Court to be tried for high treason.

2nd. I have the moral courage to show the truth of my text above, because I believe in the resurrection of life.

3rd. Brave comrades in arms who fell on that disgraced Sabbath morning, December 3rd, worthy of a better fate, and most certainly of a longer remembrance, it is in my power to drag your names from an ignoble oblivion, and vindicate the unrewarded bravery of one of yourselves! He was once my mate, the bearer of our standard, the "Southern Cross." Shot down by a murderous hand, he fell and died struggling like a man in the cause of the diggers. But he was soon forgotten. That he was buried is known by the tears of a few true friends! the place of his burial is little known, and less cared for.

Sunt tempora nostra; non mutabimur nec mutamur in illis; jam perdidi spem.

The work will be published on the 1st of December next, and given to each subscriber by the Author's own hand, on the site of the Eureka Stockade, from the rising to the setting of the sun, on the memorable third.

1. Listen to me —
   The lie, like the whirlwind, clears itself a royal road, either in town or country, through the whole face of the earth.
   The fool in his heart says, "There is no God."
   The truth, however slow, step by step, like a little child, someday, at last, finds a footpath to light.
   Then the righteous flourish like a palm tree.

———

II.

A JOVE PRINCIPIUM.

"WANTED a Governor. Apply to the People of Victoria:" that was the extraordinary advertisement, a new chum in want of employment, did meet in the usual column of *The Argus,* December 1852. Many could afford to laugh at it, the intelligent however, who had immigrated here, permanently to better his condition, was forced to rip up in his memory a certain fable of Æsop. Who would have dared *then* to warn the fattened Melbourne frogs weltering in grog, their colonial glory, against their contempt for King Log? Behold King Stork is your reward. *Tout comme chez nous.*

One remark before I start for the gold-fields. As an old European traveller I had set apart a few coppers for the poor at my landing. I had no opportunity for them. "We shall do well in this land;" was my motto. Who is going to be the first beggar? Not I! My care for the poor would have less disappointed me, if I had prepared myself against falling in the unsparing clutches of a shoal of land-sharks, who swarmed at that time the Yarra Yarra wharfs. Five pounds for landing my luggage, was the A, followed by the old colonial C, preceded by the double D. Rapacity in Australia is the alpha and omega. Yet there were no poor! a grand reflection for the serious. Adam Smith, settled the question of "the wealth of nations." The source of pauperism will be settled in Victoria by any quill-driver, who has the pluck to write the history of public-houses in the towns, and sly-grog sellers on the gold-fields.

Let us start for Ballaarat, Christmas, December, 1852. — Vide — tempore suo —

Julii Cæsaris junioris. De Campis Aureis, Australia Felix Commentaria.
For the purpose, it is now sufficient to say that I had joined a party; fixed our tent on the Canadian Flat; went up to the Camp to get our gold licence; for one pound ten shilling sterling a head we were duly licensed for one month to dig, search for, and remove gold, etc. — We wanted to drink a glass of porter to our future success, but there was no Bath Hotel at the time. — Proceeded to inspect the famous Golden Point (a sketch of which I had seen in London in the *Illustrated News*). The holes all around, three feet in diameter, and five to eight feet in depth, had been abandoned; we jumped into one, and one of my mates gave me the first lesson in "fossiking," — In less than five minutes I pounced on a little pouch — the yellow boy was all there, — my eyes were sparkling, — I felt a sensation identical to a first declaration of love in by-gone times. — *"Great works,"* at last was my bursting exclamation. In old Europe I had to take off my hat half a dozen times, and walk from east to west before I could earn one pound in the capacity of sworn interpreter, and translator of languages in the city of London. Here, I had earned double the amount in a few minutes, without crouching or crawling to Jew or Christian. Had my good angel prevailed on me to stick to that blessed Golden Point, I should have *now* to relate a very different story: the gold fever, however, got the best of my usual judgment, and I dreamt of, and pretended nothing else, than a hole choked with gold, sunk with my darling pick, and on virgin ground. — I started the hill right-hand side, ascending Canadian Gully, and safe as the Bank of England I pounced on gold — seventeen and a half ounces, depth ten feet.

III.

JUPITER TONANS.

ONE fine morning (Epiphany week), I was hard at work (excuse old chum, if I said *hard*: though my hand had been scores of times compelled in London to drop the quill through sheer fatigue, yet I never before handled a pick and shovel), I hear a rattling noise among the brush. My faithful dog, Bonaparte, would not keep under my control. "What's up?" "Your licence, mate," was the peremptory question from a six-foot fellow in blue shirt, thick boots, the face of a ruffian, armed with a carabine and fixed bayonet. The old "all right" being exchanged, I lost sight of that specimen of colonial brutedom and his similars, called, as I then learned, "traps" and "troopers." I left off work, and was unable to do a stroke more that day.

"I came, then, 16,000 miles in vain to get away from the law of the sword!" was my sad reflection. My sorrow was not mitigated by my mates and neighbours informing me, that Australia was a penal settlement. Inveterate murderers, audacious burglars, bloodthirsty bushrangers, were the ruling triumvirate, the scour of old Europe, called Vandemonians, in this bullock-drivers' land. Of course I felt tamed, and felt less angry, at the following search for licence. At the latter end of the month, one hundred and seventy-seven pounds troy, in two superb masses of gold, were discovered at the depth of sixty feet, on the hill opposite where I was working. The talk was soon Vulcanish through the land. Canadian Gully was as rich in *lumps* as other gold-fields are in *dust*. Diggers, whom the gold fever had rendered stark blind, so as to desert Ballaarat for Mount Alexander and Bendigo, now returned as ravens to the old spot; and towards the end of February, '53, Canadian Gully was in its full glory.

———————

IV.

INCIPIT LAMENTATIO.

THE search for licences, or "the traps are out to-day" — their name at the time — happened once a month. The strong population now on this gold-field had perhaps rendered it necessary twice a month. Only in October, I recollect they had come out three times. Yet, "the traps are out" was annoying, but not exasperating. Not exasperating, because John Bull, *ab initio et ante secula,* was born for law, order, and safe money-making on land and sea. They were annoying, because, said John, not that he likes his money more than his belly, but he hates the bayonet: I mean, of course, he does not want to be bullied with the bayonet. To this honest grumbling of John, the drunkard, that is the lazy, which make the incapables, joined their cant, and the Vandemonians pulled up with wonted audacity. In a word, the thirty shillings a month for the gold licence became a nuisance.

A public meeting was announced on Bakery-hill. It was in November, 1853. Four hundred diggers were present. I recollect I heard a "Doctor Carr" poking about among the heaps of empty bottles all round the Camp, and asked who paid for the good stuff that was in them, and whither was it gone. Of course, Doctor Carr did not mention, that one of those bottles, corked and sealed with the "Crown," was forced open with Mr. *Hetherington's* corkscrew; and that said Dr. Carr had then to confess that the bottle aforesaid contained a nobbler some £250 worth for himself. Great works already at Toorak. *Tout cela soit dit en passant.* Mr. Hetherington, then a storekeeper on the Ballaarat Flat, and now of the Clarendon Hotel, Ballaarat Township, is a living witness. For the fun of the thing, I spoke a few words which merited me a compliment from the practitioner, who also honoured me with a private precious piece of information — *"Nous allons bientot avoir la Republique Australienne! Signore." "Quelle farce! repondis je."* The specimen of man before me impressed me with such a decided opinion of his ability for destroying sugarsticks, that at once I gave him credit as the founder of a republic for babies to suck their thumbs.

In short, here dates the Victorian system of "memorialising." The diggers of Ballaarat sympathised with those of Bendigo in their common grievances, and prayed the governor that the gold licence be reduced to thirty shillings a month. There was further a great waste of yabber-yabber about the diggers not being represented in the Legislative Council, and a deal of fustian was spun against the squatters. I understood very little of those matters at the time: the shoe had not pinched my toe yet.

4

Every one returned to his work; some perhaps not very peacefully, on account of a nobbler or two over the usual allowance.

V.

RISUM TENEATIS AMICI.

I RECOLLECT towards this time I followed the mob to Magpie Gully. It was a digger's life. Hard work by day, blazing fire in the evening, and sound sleep by night at the music of drunken quarrels all around, far and near. I had marked my claim in accordance with the run of the ranges, and safe as the Bank of England I bottomed on gold. No search for licence ever took place. What's the matter? Oh, the diggers of Bendigo, by sheer moral force, in the shape of some ten thousand in a mob, had inspired with better sense the red-tape there and somewhere else, so I took out my licence at the reasonable rate of two pounds for three months, my contribution for the support of gold-lace. So far so good. I had no fault to find with our governor Joseph Latrobe, Esquire; nor do I believe that the diggers cared about anything else from him. Was it then his being an esquire that brought his administration into contempt? The fact is, a clap of "The Thunder" from Printing House-square boomed on the tympanum of my ear. We diggers got the gracious title of "vagabonds," and our massa "Joe," for his pains to keep friends with us, was put down "an incapable;" all for the honour of British rule, of course.

"Wanted a Governor," was now no longer a dummy in *The Argus*; but, unhappily, no application was made to the people of Victoria.

Give a dog a bad name — and the old proverb holds good even at the antipodes. My trampings are now transcribed from my diary.

With the hot winds whirled in the Vandemonian rush to the Ballaarat Flat. My hole was next to the one which was jumped by the Eureka mob, and where one man was murdered in the row. At sixty-five feet we got on a blasted log of a gum-tree that had been mouldering there under a curse, since the times of Noah! The whole flat turned out an imperial shicer. (You did not sink deep enough, Signore Editor.) Slabs that had cost us some eight pounds a hundred would not fetch, afterwards, one pound. We left them to sweat freely in the hole; and all the mob got on the fuddle. My mate and myself thought we had been long enough together, and got asunder for a change. I was soon on the tramp again. Bryant's Ranges was the go of the day, and I started thither accordingly. December, 1853. Oh, Lord! what a pack of ragamuffins over that way! I got acquainted with the German party who found out the Tarrangower den; shaped my hole like a bathing tub, and dropped "on it" right smart. Paid two pounds to cart one load down the Loddon, and left two more loads of washing stuff, snug and wet with the sweat of my brow, over the hole. Got twenty-eight pennyweights out of the load. Went back the third day, brisk and healthy, to cart down the other two loads. Washing stuff! gone: hole! gone: the gully itself! gone: the whole face of it had been clean shaved. Never mind, go ahead again. Got another claim on the surface-hill. No search for licence: thank God, had none. Nasty, sneaky, cheeky little things of flies got into my eyes: could see no more, no ways. Mud water one shilling a bucket! Got the dysentery; very bad. Thought, one night, to reef the yards and drop the anchor. Got on a better tack though. Promenaded up to the famous Bendigo. Had no particular objection to Celestials there, but had no particular taste for their tartaric water. Made up my mind to remember my days of innocence, and turned shepherd. Fine landscape this run on the Loddon: almost a match for Bella Italia; but there are too many mosquitoes. Dreamt, one day, I was drinking a tumbler of Loddon wine; and asserted that Providence was the same also in the south. It was a dream. The lands lay waste and desolate: not by nature; oh no; by hand of man. Bathing in these Loddon water-holes, superb. Tea out of this Loddon water magnificent. In spite of these horrible hot winds, this water is always fresh and delicious: how kind is Providence! One night lost the whole blessed lot of my flock. Myself, the shepherd, did not know, in the name of heavens, which way to turn. Got among the blacks, the whole Tarrang tribe in corrobory. Lord, what a rum sight for an old European traveller. Found natives very humane, though. My sheep right again, only the wild dogs had given them a good shake. Was satisfied that the Messiah the Jews are looking for will not be born in this bullock-drivers' land; any how, the angels won't announce the happy event of his birth to the shepherds. No more truck with sheep, and went to live with the blacks for a variation. Picked up, pretty soon, bits of their yabber-yabber. For a couple of years had tasted no fish; now I pounced on a couple of frogs, every couple of minutes. Thought their "lubras" ugly enough; not so, however, the slender arms and small hands of their young girls, though the fingers be rather too long.

That will do now, in as much as the end of the story is this: That portion in my brains called "acquisitiveness" got the gold-fever again, and I started for old Ballaarat.

VI.

SUA CUIQUE VOLUNTAS.

I WAS really delighted to see the old spot once more; Easter, 1854. I do not mean any offence to my fellow-diggers elsewhere; it struck me very forcibly, however, that our Ballaarat men look by far more decent, and our storekeepers, or grog-sellers if you like, undoubtedly more respectable.

Of a constitution not necessarily savage, I did not fail to observe that the fair ones had ventured now on a larger scale to trust their virtue among us vagabonds, and on a hot-wind day, I patronized of course some refreshment room.

I met my old mate, and we determined to try the old game; but this time on the old principle of *labor omnia vincit* — I pitched my tent right in the bush, and prophesied, that from my door I would see the golden hole in the gully below.

I spoke the truth, and such is the case this very day. Feast of the Assumption, 1855: — What sad events, however, were destined to pass exactly before the very

6

door of my tent! Who could have told me on that Easter Sunday, that the unknown hill which I had chosen for my rest, would soon be called the Massacre Hill! That next Christmas, my mate would lie in the grave, somewhere forgotten; and I in the gaol! the rope round my neck!!

Let us keep in good spirits, good reader, we shall soon have to weep together enough.

Gravel Pits, famous for its strong muster of golden holes, and blasting shicers, was too deep for me. The old Eureka was itself again. The jewellers' shops, which threatened to exhaust themselves in Canadian Gully, were again the talk of the day: and the Eureka gold dust was finer, purer, brighter, immensely darling. The unfaithful truants who had rushed to Bryant's Ranges, to knock their heads against blocks of granite, now hastened for the *third* time to the old spot, Ballaarat, determined to stick to it for life or death. English, German, and Scotch diggers, worked generally on the Gravel Pits; the Irish had their stronghold on the Eureka. The Americans fraternised with all the wide-awake, *ubi caro, ibi vultures.*

Here begins as a profession the precious game of "shepherding," or keeping claims in reserve; that is the digger turning squatter. And, as this happened under the reign of a gracious gold commissioner, so I am brought to speak of the gold licence again. First I will place the man before my reader, though.

Get a tolerable young pig, make it stand on his hind legs, put on its head a cap trimmed with gold-lace, whitewash its snout, and there you have the ass in the form of a pig; I mean to say a "man;" with this privilege, that he possesses in his head the brains of both the above-mentioned brutes.

VII.

LUDI BALLAARATENSES.

EUREKA was advancing fast to glory. Each day, and not seldom twice a day, the gutter gammoned and humbugged all us "vagabonds" so deucedly, that the rush to secure a claim "dead on it" rose to the standard of "Eureka style;" that is, "Ring, ring," was the yell from some hundred human dogs, and soon hill and flat poured out all spare hands to thicken the "ring."

By this time, two covies — one of them generally an Irishman — had stripped to their middle, and were "shaping" for a round or two. A broken nose, with the desired accomplishment of a pair of black eyes; and in all cases, when manageable, a good smash in the regions either of the teeth, or of the ribs — both, if possible, preferred — was supposed to improve the transaction so much, that, what with the tooth dropping, or the rib cracking, or both, as aforesaid, it was considered "settled." Thus originated the special title of "rowdy mob," or Tipperary, in reference to the Irish. Let us have the title clear.

The "shepherding," that is the squatting by one man — women and children had not got hold of this *Dolce far niente* yet — the ground allotted by law to four men; and the astuteness of our primitive shepherds having found it cheap and profitable to have each claim visibly separated from the other by some twenty-feet wall, which was mutually agreed upon by themselves alone, to call it "spare ground," was now a grown-up institution. Hence, whenever the gutter, 120 feet below, took it into its head to bestir and hook it, the faithful shepherds would not rest until they were sure to snore in peace a foot and a half under ground from the surface, and six score feet from "bang on the gutter."

This Ballaarat dodge would have been innocent enough, were it not for "Young Ireland," who, having fixed head-quarters on the Eureka, was therefore accused of monopolising the concern. Now, suppose Paddy wanted to relish a "tip," that is, a drop of gin on the sly, then Scotty, who had just gulped down his "toddy," which was a drop of auld whisky, would take upon himself the selfish trouble to sink six inches more in Paddy's hole, which feat was called "jumping;" and thus, broken noses, and other accomplishments, as aforesaid, grew in proportion to tips, and "toddy" drunk on the sly.

I frequently saw horrid scenes of blood; but I was now an old chum, and therefore knew what was what in colonial life.

I had a Cameleon for a neighbour, who, in the garb of an Irishman, flung his three half-shovels out of a hole on the hill punctually every morning, and that was his work before breakfast. Then, a red shirt on his back, and a red cap on his head, he would, in the subsequent hour, give evidence of his scorning to be lazy by putting down some three inches deeper another hole below in the gully. "Full stop;" he must have a "blow," but the d——d things — his matches — had got damp, and so in a rage he must hasten to his tent to light the pipe; that is, to put on the Yankee garb and complete his forenoon work in a third hole of his, whose depth and shape recommended him as a first rate grave-digger.

And what has all this bosh to do with the Eureka Stockade?

———

VIII.

FIAT JUSTITIA, RUAT CŒLUM.

AS an old Ballaarat hand, I hereby assert, that much of the odium of the mining community against red-tape, arose from the accursed practice of jumping.

One fact from the "stubborn-things" store. The Eureka gutter was fast progressing down hill towards the Eureka gully. A party of Britishers had two claims; the one, on the slope of the hill, was bottomed on heavy gold; the other, some four claims from it, and parallel with the range, was some ninety feet deep, and was worked by day only, by three men: a fourth man would now and then bring a set of trimmed slabs from the first hole aforesaid, where he was the principal "chips." There was a Judas Iscariot among the party. One fine morning, a hole was bottomed down the gully, and proved a scheisser. A rush, Eureka style, was the conseqence; and it was pretended now that the gutter would keep with the ranges, towards the Catholic church.

A party of Yankees, with revolvers and Mexican knives — the garb of "bouncers" in those days — jumped the second hole of the Britishers, dismantled the windlass, and Godamn'd as fast as the Britishers cursed in the colonial style. The excitement was awful. Commissioner Rede was fetched to settle the dispute. An absurd and unjust regulation was then the law; no party was allowed to have an interest in two claims at one and the same time, which was called "owning two claims." The Yankees carried the day. I, a living witness, do assert that, from that day, there was a "down" on the name of Rede.

For the commissioners, this jumping business was by no means an agreeable job. They were fetched to the spot: a mob would soon collect round the disputed claim; and for "fair play," it required the wisdom of Solomon, because the parties concerned set the same price on their dispute, as the two harlots on the living child.

I. The conflicting evidence, in consequence of hard swearing, prompted by gold-thirst, the most horrible demon that depraves the human heart, even a naturally honest heart. — II. The incomprehensible, unsettled, impracticable ordinances for the abominable management of the gold-fields; which ordinances, left to the discretion — that is, the caprice; and to the good sense — that is, the motto, *"odi profanum vulgus et arceo;"* and to the best judgment — that is the proverbial incapability of all aristocractical red-tape, HOW TO RULE US VAGABONDS. Both those reasons, I say, must make even the most hardened bibber of Toorak small-beer acknowledge and confess, that the perfidious mistake at head-quarters was, their persisting to make the following Belgravian *billet-doux* the *"sine qua non"* recommendation for gold-lace on Ballaarat (*at the time*): —

(ADDRESS)
"To the Victorian Board of Small Beer,
"Toorak (somewhere in Australasia, *i.e.*, Australia Felix — inquire from the natives, reported to be of black-skin, at the southern end of the globe.)
"Belgravia,
"First year of the royal projecting the
"Great Exhibition, Hyde Park.

"Lady STARVESEMPSTRESS, great-grand-niece of His Grace the Duke Of CURRY-POWDER, begs to introduce to FORTYSHILLING TAKEHIMAWAY, Esquire, of Toorak, see address, her brother-in-law, POLLIPUSS, WATERLOOBOLTER, tenth son of the venerable Prebendary of North and South Palaver, Canon of St. Sebastopol in the east, and Rector of Allblessedfools, West End — URGENT."

In justice, however, to Master Waterloobolter, candidate for gold-lace, it must not be omitted that he is a Piccadilly young sprat, and so at Julien's giant bal-masque, was ever gracious to the lady of his love.

"Miss Smartdeuce, may I beg the honour of your hand for the next waltz? surely after a round or two you will relish your champagne."

"Yes," with a smothered "dear," was the sigh-drawn reply.

Who has the power to roar the command, "Thus far shalt thou go, and no further," to the flood of tears from forlorn Smartdeuce, when her soft

Waterloobolter bolted for the gold-fields of Australia Felix.

To be serious. How could any candid mind otherwise explain the honest boldness of eight out of nine members of the first Local Court, Ballaarat, who, one and all, I do not say dared, but I say called upon their fellow miners to come forward to a public meeting on the old spot, Bakery-hill. September, Saturday, 30th, 1855. Said members had already settled at that time 201 disputes, and given their judgment, involving some half a million sterling altogether, for all what they knew, and yet not one miner rose one finger against them, when they imperatively desired to know whether they had done their duty and still possessed the confidence of their fellow diggers! They (said members) are practical men, of our own adopted class, elected by ourselves from among ourselves, to sit as arbitrators of our disputes, and our representatives at the Local Court. That's the key, for any future Brougham, for the history of the Local Courts on the gold-fields.

It has fallen to my lot, however, to put the Eureka Stockade on record; and, from the following "Joe"-chapter must begin any proper history of that disgracefully memorable event.

IX.

ABYSSUS. ABYSSUM INVOCAT.

"JOE, Joe!" No one in the world can properly understand and describe this shouting of "Joe," unless he were on this El Dorado of Ballaarat at the time.

It was a horrible day, plagued by the hot winds. A blast of the hurricane winding through gravel pits whirled towards the Eureka this shouting of "Joe." It was the howl of a wolf for the shepherds, who bolted at once towards the bush: it was the yell of bull-dogs for the fossikers who floundered among the deep holes, and thus dodged the hounds: it was a scarecrow for the miners, who now scrambled down to the deep, and left a licensed mate or two at the windlass. By this time, a regiment of troopers, in full gallop, had besieged the whole Eureka, and the traps under their protection ventured among the holes. An attempt to give an idea of such disgusting and contemptible campaigns for the search of licences is really odious to an honest man. Some of the traps were civil enough; aye, they felt the shame of their duty; but there were among them devils at heart, who enjoyed the fun, because their cupidity could not bear the sight of the zig-zag uninterrupted muster of piles of rich-looking washing stuff, and the envy which blinded their eyes prevented them from taking into account the overwhelming number of shicers close by, round about, all along. Hence they looked upon the ragged muddy blue shirt as an object of their contempt.

Are diggers dogs or savages, that they are to be hunted on the diggings, commanded, in Pellissier's African style, to come out of their holes, and summoned from their tents by these hounds of the executive? Is the garb of a digger a mark of inferiority? "*In sudore vultus tue vesceris panem*"[1] is then an infamy now-a-days!

Give us facts, and spare us your bosh, says my good reader. — Very well.

I, CARBONI RAFFAELLO, da Roma, and late of No. 4, Castle-court, Cornhill, City of London, had my rattling "Jenny Lind" (the cradle) at a water-hole down the Eureka Gully. Must stop my work to shew my licence. "All right." I had then to go a quarter of a mile up the hill to my hole, and fetch the washing stuff. There again — "Got your licence?" "All serene, governor." On crossing the holes, up to the knees in mullock, and loaded like a dromedary, "Got your licence?" was again the cheer-up from a third trooper or trap. Now, what answer would you have given, sir?

I assert, as a matter of fact, that I was often compelled to produce my licence twice at each and the same licence hunt. Any one who knows me personally, will readily believe that the accursed game worried me to death.

1. "In the sweat of thy brow thou shalt eat bread."

X.

JAM NON ESTIS HOSPITES ET ADVENÆ.

IT is to the purpose to say a few words more on the licence-hunting, and have done with it. Light your pipe, good reader, you have to blow hard.

Our red-tape, generally obtuse and arrogant, this once got rid of the usual conceit in all things, and had to acknowledge that the digger who remained quietly at his work, always possessed his licence. Hence the troopers were despatched like bloodhounds, in all directions, to beat the bush; and the traps who had a more confined scent, creeped and crawled among the holes, and sneaked into the sly-grog tents round about, in search of the swarming unlicensed game. In a word, it was a regular hunt. Any one who in Old England went fox-hunting, can understand pretty well, the detestable sport we had then on the gold-fields of Victoria. Did any trooper succeed in catching any of the "vagabonds" in the bush, he would by the threat of his sword, confine him round a big gum-tree; and when all the successful troopers had done the same feat, they took their prisoners down the gully, where was the grand depot, because the traps were generally more successful. The commissioner would then pick up one pound, two pounds, or five pounds, in the way of bail, from any digger that could afford it, or had friends to do so, and then order the whole pack of the penniless and friendless to the lock-up in the camp. *I am a living eye-witness, and challenge contradiction.*

This job of explaining a licence-hunt is really so disgusting to me, that I prefer to close it with the following document from my subsequently goal-bird mate, then reporter of the *Ballaarat Times:* —

Police Court, Tuesday, October 24th.
HUNTING THE DIGGER. — Five of these fellows were fined in the

mitigated trifle of £5, for being without licences. The nicest thing imaginable is to see one of these clumsy fellows with great beards, shaggy hair, and oh! such nasty rough hands, stand before a fine gentleman on the bench with hands of shiny whiteness, and the colour of whose cambric rivals the Alpine snow. There the clumsy fellow stands, faltering out an awkward apology, "my licence is only just expired, sir — I've only been one day from town, sir — I have no money, sir, for I had to borrow half a bag of flour the other day, for my wife and children." Ahem, says his worship, the law makes no distinctions — fined £5. Now our reporter enjoys this exceedingly, for he is sometimes scarce of news; and from a strange aberration of intellect, with which, poor fellow, he is afflicted, has sometimes, no news at all for us; but he is sure of not being *dead beat* at any time, for digger-hunting is a standing case at the police office, and our reporter is growing so precocious with long practice, that he can tell the number of diggers fined every morning, without going to that sanctuary at all. — *Ballaarat Times,* Saturday, October 28, 1854.

XI.

SALVUM FAC POPULUM TUUM DOMINE.

THE more the pity — I have not done yet with the accursed gold licence. I must prevail on myself to keep cooler and in good temper.

Two questions will certainly be put to me: —

1st. Did the camp officials give out the licence to the digger at the place of his work, whenever required, without compelling him to leave off work, and renew his licence at the camp?

2nd. It was only one day in each month that there was a search for licences, was it not? Why therefore did not the diggers make it a half-holiday on the old ground, that "all work and no play, makes Jack a dull boy."

The first question is a foolish one, from any fellow-colonist who knows our silver and gold lace; and is a wicked one, from any digger who was on Ballaarat at the time.

"Fellah" gave the proper answer through the *Ballaarat Times,* October 14th; — here it is: —

To the Editor of the *Ballaarat Times,* October 14, 1854.

Sir,

Permit me to call your attention to the miserable accommodation provided for the miner, who may have occasion to go to the Camp to take out a licence. Surely, with the thousands of pounds that have been expended in government buildings, a little better accommodation might be afforded to the well disposed digger, who is willing to pay the odious tax demanded of him by government, and not be compelled to stand in the rain or sun, or treated as if the "distinguished government official" feared that the digger was a thing that would contaminate him by a closer proximity; so the "fellah" is kept by a wooden rail from approaching within a couple of yards of the tent. In consequence of so many persons mistaking the licence-office for the commissioner's water-closet, a placard has been placed over the door.

I am, Sir, yours &c.,

FELLAH DIGGER,

Who had to walk a few miles to pay away the money he had worked hard for, and was kept a few hours standing by a rail — not "sitting on a rail, Mary."

Now I mean to tackle in right earnest with the second question, provided I can keep in sufficiently good temper.

On the morning of Thursday, the 22nd June, in the year of Grace, One thousand eight hundred and fifty-four,

His Excellency Sir CHARLES HOTHAM,

Knight Commander of the Most Noble Military Order of the Bath, landed on the shores of this fair province, as its Lieutenant-Governor, the chosen and commissioned representative of Her Most Gracious Majesty, the QUEEN! "Never (writes the Melbourne historian of that day) never in the history of public ovations, was welcome more hearty, never did stranger meet with warmer welcome, on the threshold of a new home:

VICTORIA WELCOMES VICTORIA'S CHOICE,

was the Melbourne proclamation.

The following is transcribed from my diary: —

Saturday, August 26th, 1854: His Excellency dashed in among us "vagabonds" on a sudden, at about five o'clock p.m., and inspected a shaft immediately behind the Ballaarat Dining Rooms, Gravel-pits. A mob soon collected round the hole; we were respectful, and there was no "joeing." On His Excellency's return to the camp, the miners busily employed themselves in laying down slabs to facilitate his progress. I was among the zealous ones who improvised this shabby foot-path. What a lack! we were all of us as cheerful as fighting-cocks. — A crab-hole being in the way, our Big-Larry actually pounced on Lady Hotham, and lifting her up in his arms, eloped with her ladyship safely across, amid hearty peals of laughter, however colonial they may have been. — Now, Big Larry kept the crowd from annoying the couple, by properly laying about him with a switch all along the road.

His Excellency was hailed with three-times-three, and was proclaimed on the Camp, now invaded by some five hundred blue shirts, the "Diggers' Charley."

His Excellency addressed us miners as follows: — "Diggers I feel delighted with your reception — I shall not neglect your interests and welfare — again I thank you."

It was a short but smart speech we had heard elsewhere, he was not fond of "twaddle," which I suppose meant "bosh." After giving three hearty cheers, old Briton's style to "Charley," the crowd dispersed to drink a nobbler to his health and success. I do so this very moment. Eureka, under my snug tent on the hill, August 26, 1854. C.R."

Within six short months, *five thousand* citizens of Melbourne, receive the name of this applauded ruler with a loud and prolonged outburst of indignation!

Some twenty Ballaarat miners lie in the grave, weltering in their gore! double that number are bleeding from bayonet wounds; thirteen more have the rope round their necks, and two more of their leading men are priced four hundred pounds for their body or carcase.

Tout cela, n'est pas precisement comme chez nous, pas vrai?

Please, give me a dozen puffs at my black-stump, and then I will proceed to the next chapter.

XII.

SIFFICIT DIEI SUA VEXATIO.

EITHER this chapter must be very short, or I had better give it up without starting it at all.

Up to the middle of September, 1854, the search for licences happened once a month; at most twice: perhaps once a week on the Gravel Pits, owing to the near neighbourhood of the Camp. Now, licence-hunting became the order of the day. Twice a week on every line; and the more the diggers felt annoyed at it, the more our Camp officials persisted in goading us, to render our yoke palatable by habit. I assert, as an eye-witness and a sufferer, that both in October and November, when the weather allowed it, the Camp rode out for the hunt every alternate day. True, one day they would hunt their game on Gravel-pits, another day, they pounced on the foxes of the Eureka; and a third day, on the Red-hill: but, though working on different leads, are we not all fellow diggers? Did not several of us meet again in the evening, under the same tent, belonging to the same party? It is useless to ask further questions.

Towards the latter end of October and the beginning of November we had such a set of scoundrels camped among us, in the shape of troopers and traps, that I had better shut up this chapter at once, or else whirl the whole manuscript bang down a shicer.

"Hold hard, though, take your time, old man: don't let your Roman blood hurry you off like the hurricane, and thus damage the merits of your case. Answer this question first," says my good reader.

"If it be a fair one, I will."

"Was, then, the obnoxious mode of collecting the tax the sole cause of discontent: or was the tax itself (two pounds for three months) objected to at the same time?"

"I think *the practical miner,* who had been *hard at work, night and day,* for the last four or six months, and, after all, had just bottomed a shicer, objected to the tax itself, because he could not possibly afford to pay it. And was it not atrocious to confine this man in the lousy lock-up at the Camp, because he had no luck?"

Allow me, now, in return, to put a very important question, of the old Roman stamp, *Cui bono?* that is, *Where did our licence money go to?* That's a nut which will be positively cracked by-and-bye.

XIII.

UBI CARO, IBI VULTURES.

ONE morning, I woke all on a sudden. — What's up? A troop of horse was galloping exactly towards my tent, and I could hear the tramping of a band of traps. I got out of the stretcher, and hastened out of my tent. All the neighbours, in night-caps and unmentionables, were groping round the tents, to inquire what was the matter. It was not yet day-light. There was a sly-grog seller at the top of the hill; close to his store he had a small tent, crammed with brandy cases and other grog, newly come up from town. There must have been a spy, who had scented such valuable game.

The Commissioner asked the storekeeper, who by this time was at the door of his store: "Whose tent is that?" indicating the small one in question.

"I don't know," was the answer.

"Who lives in it? who owns it? is anybody in?" asked the Commissioner.

"An old man owns it, but he is gone to town on business, and left it to the care of his mate who is on the night-shift," replied the storekeeper.

"I won't peck up that chaff of yours, sir. Halloo! who is in? Open the tent;" shouted the Commissioner.

No answer.

"I say, cut down this tent, and we'll see who is in;" was the order of the Commissioner to two ruffianly looking troopers.

No sooner said than done; and the little tent was ripped up by their swords. A government cart was, of course, ready in the gully below, and in less than five minutes the whole stock of grog, some two hundred pounds sterling worth, or five hundred pounds worth in nobblers, was carted up to the Camp, before the teeth of some hundreds of diggers, who had now collected round about. We cried "Shame! shame!" sulkily enough, but we did not interfere; first, because the store had already annoyed us often enough during the long winter nights; second, because the plunderers were such Vandemonian-looking traps and troopers, that we were not encouraged to say much, because it would have been of no use.

As soon, however, as the sun was up, and all hands were going to work, the occurrence not only increased the discontent that had been brewing fast enough already, but it rose to excitement; and such a state of exasperated feelings, however vented in the shouting of "Joe," did certainly not prepare the Eureka boys to submit with patience to a licence-hunt in the course of the day.

First and foremost: it is impossible to prevent the sale of spirits on the diggings; and not any laws, fines, or punishment the government may impose on the dealers or consumers can have any effect towards putting a stop to sly-grog selling. A miner working, as during the past winter, in wet and cold, *must* and *will* have his nobbler occasionally; and very necessary, too, I think. No matter what the cost, he will have it; and it cannot be dispensed with, if he wish to preserve his health: he won't go to the Charley Napier Hotel, when he can get his nobbler near-handy, and thereby give a lift to Pat or Scotty.

Secondly: I hereby assert that the breed of spies in this colony prospered by this sly-grog selling. "We want money," says some of the "paternals" at Toorak.

"Oh! well, then," replies another at Ballaarat, "come down on a few storekeepers and unlicensed miners and raise the wind. We can manage a thousand or two that way. Let the blood-hounds on the scent, and it is done."

And so a scoundrel, in the disguise of an honest man, takes with him another worse devil than himself, and goes round like a roaring lion, seeking what he may devour.

If I had half the fifty pounds fine inflicted on sly-grog sellers, and five pounds fine on unlicensed diggers, raised on Ballaarat at this time, I think my fellow-colonists would bow their heads before me. Great works!

Thirdly: An act of silver and gold lace humanity was going the rounds of our holes, above and below.

A person is found in an insensible state, caused by loss of blood, having fallen, by accident, on a broken bottle and cut an artery in his head. He is conveyed to the Camp hospital.

After some few hours, because he raves from loss of blood, and at a time when he requires the closest attention, he is unceremoniously carried into the common lock-up, and there left, it is said, for ten hours, lying on the floor, without any attention being paid to his condition by the hospital authorities, and then it was only by repeated representations of his sinking state, to other officials, that he was conveyed to the hospital, where *he expired in two hours afterwards!*

"Below!"

"Haloo!"

"Jim; the miners of Ballaarat *demand* an investigation."

"And *they must have it,* Joe."

Such was the scene in those days, performed at every shaft, in Gravel-pits, as well as on the Eureka.

XIV.

FLAGITATUR VULCANO SI FULMINA PARATA.

HERE is a short *resumé* of events which led to the popular demonstration on Tuesday, October 17th, 1854.

Two men, old friends, named Scobie and Martin, after many years' separation, happened to meet each other in Ballaarat. Joy at the meeting, led them to indulge in a wee drop for "Auld lang Syne." In this state of happy feeling, they call at the Eureka Hotel, on their way home, intending to have a finishing glass. They knock at the door, and are refused admittance, very properly, on account of their drunkenness. They leave, and proceed on their way, not, perhaps, without the usual colonial salutations. At about fifty yards from the hotel, they hear a noise behind them, and retrace their steps. They are met by persons, unknown, who inflict blows on them, which render one insensible and the other lifeless.

A coroner's inquest was held on the body, the verdict of which was, "that deceased had died from injuries inflicted by persons unknown;" but public feeling seemed to point to Mr. Bentley, the proprietor of the Eureka Hotel; who, together with his wife and another party, were charged with the murder, tried at the police court, and acquitted.

The friends of deceased, considering that both the inquest and the trial were unfairly conducted, agreed to meet on Tuesday, October 17th, on the spot where the man was murdered, and devise measures to discover the guilty parties, and to bring them to justice.

Accordingly, at an early hour, the hill on which is situated the Eureka Hotel was thronged by thousands; so great was the excitement.

THOMAS KENNEDY, was naturally enough the lion of the day. A thick head, bold, but bald, the consequence perhaps not of his dissipation; but of his worry in by gone days. His merit consists in the possession of the chartist slang; hence his cleverness in spinning a yarn, never to the purpose, but blathered with long phrases and bubbling with cant. He took up the cause of the diggers, not so much for the evaporation of his gaseous heroism, as eternally to hammer on the unfortunate death of his country-man Scobie, for the sake of "auld lang syne."

When pressed by the example of others to burn his licence, at the subsequent monster meeting, he had none to burn, because he had a wife and four children dependent on him for support, and therefore I do not know what to say further.

These and other resolutions were carried unanimously: —

"That this meeting, not being satisfied with the manner in which the proceedings connected with the death of the late James Scobie, have been conducted, either by the magistrates or by the coroner, pledges itself to use every lawful means to have the case brought before other, and more competent authorities.

"That this meeting deems it necessary to collect subscriptions for the purpose of offering a reward for the conviction of the murderers, and defraying all other expenses connected with the prosecution of the case."

XV.

NAM TUA RES AGITUR, PARIES CUM PROXIMUS ARDET.

THE one pervading opinion among the multitude of miners and others who had been attracted thither, appeared to be, that Bentley was the murderer; and loud were the cries, the hooting, and groans against him. It would appear that the Camp authorities contemplated some little disturbance, and consequently, all the available force of police and mounted troopers were on guard at the hotel and made a very injudicious display of their strength. Not only did they follow, but ride through, the crowd of people at the meeting; and it is to this display of their strength that must be attributed the fire, and other outbursts of indignation. Miners who have stood the working of a Canadian or Gravel-pit shicer, scorn danger in any form.

The crowd, excessively irritated on seeing the large display of the hated police force, now began to shout and and yell. Presently, a stone came from the mass, and passing near the head of one of the officials, broke a pane of glass in one of the windows of the hotel. The sound of the falling glass appeared to act like magic on the multitude; and bottles, stones, sticks, and other missiles, were speedily put in requisition to demolish the windows, until not a single pane was left entire, while every one that was broken drew a cheer from the crowd. The police, all this time, were riding round and round the hotel, but did not take any vigorous measures to deter the people from the sport they appeared to enjoy so much. The crowd advance nearer — near enough to use sticks to beat in the casements. They make an entrance, and, in a moment, furniture, wearing apparel, bedding, drapery, are tossed out of the windows; curtains, sheets, etc., are thrown in the air, frightening the horses of the troopers, who have enough to do to keep their saddles; the weather-boards are ripped off the side of the house, and sent spinning in the air. A real Californian takes particular care of, and delights in smashing the crockery.

Mr. Rede, the resident Commissioner, arrives, and endeavours to pacify the people by speechifying; but it will not do. He mounts the sill of where was once a window, and gesticulates to the crowd to hear him. An egg is thrown from behind a tent opposite, and narrowly misses his face, but breaks on the wall of the house close to him. The Commissioner becomes excited, and orders the troopers to take the man in charge; but no trooper appears to relish the business.

A cry of "Fire!" is raised; a horse shies and causes commotion. Smoke is seen to issue from one of the rooms of the ground-floor. The police extinguish it; and an attempt is made to form a cordon round the building. But it is too late. Whilst the front of the hotel occupies the attention of the majority of the crowd, a few are pulling down the back premises.

Mr. Rede sends for the detachment of the gallant 40th, now stationed on Ballaarat.

A shout is raised: — "The 40th are coming."

"Don't *illuminate* till they come."

"They shall see the sight."

"Wait till they come."

Smash go the large lamps in front of the hotel. The troopers ride round and caracole their horses.

"Where's the red-coats?"

"There they come, yonder up the hill!"

"Hurrah! three cheers."

The 40th arrive; they form into line in front of the hotel, swords drawn. "Hurrah! boys! no use waiting any longer." — "Down she comes." The bowling alley is on fire. — Police try to extinguish the flames — rather too warm. — It's too late. — The hotel is on fire at the back corner; nothing can save it. — "Hip, hip hurrah!" is the universal shout.

I had opportunities enough to observe in London, that a characteristic of the British race is to make fun of the calamity of fire, hence I did not wonder, how they enjoyed this, their real sport on the occasion.

A gale of wind, which blowed at this exact time, announcing the hurricane that soon followed, was the principal helper to the devouring of the building, by blowing in the direction most favourable to the purpose.

The red-coats wheel about, and return to the Camp. Look out! the roof of the back part of the hotel, falls in! "Hurrah! boys, here's the porter and ale with the chill off."

Bottles are handed out burning hot — the necks of two bottles are knocked together! — Contents drunk in colonial style. — Look out! the roof, sides and all fall in! — An enormous mass of flame and smoke arises with a roaring sound. — Sparks are carried far, far into the air, and what was once the *Eureka Hotel*, is now a mass of burning embers!

The entire diggings, in a state of extreme excitement. — The diggers are lords and masters of Ballaarat; and the prestige of the Camp is gone for ever.

XVI.

LOQUAR IN AMARITUDINE ANIMÆ MEÆ.

NOW my peace of mind being destroyed, I had recourse to the free British press, for information, wishing to hear what they said in Melbourne. At this time the *Morning Herald* was in good demand; but the *Geelong Advertiser* had the sway on the gold-fields. Geelong had a rattling correspondent on Ballaarat, who helped to hasten the movement fast enough. As I did not know this correspondent of the *Geelong Advertiser* personally, so I can only guess at his frame of mind. I should say the following ingredients entered into the factory of his ideas: —

1st. The land is the Lord's and all therein; but man must earn his bread by the sweat of his brow. Therefore, in the battle of life, every man must fight his way on the old ground, "help yourself and God will help you."

2nd. In olden times, wherever there was a Roman there was life. In our times, wherever there is a Britain there is trade, and trade is life. But with the lazy, — who, either proud or mean, is always an incapable, because generally he is a drunkard, and therefore a beggar, there is no possible barter; and, inasmuch as man does not live on bread alone, for a fried sole is a nice thing for breakfast, so also it must be confessed that the loaves and fishes do not condescend to jump into one's mouth all dressed as they ought to be. Therefore — and this is the zenith of the *Geelong Advertiser's* practical correspondent — be not perplexed, if the loaves and fishes wont pop fast enough into your mouth particularly; let Mahomed's example be instantly followed: go yourself to the loaves and fishes, and you will actually find that they are subject to the same laws of matter and motion as everything else on earth.

3rd. The application. For what did any one emigrate to this colony? To sweat more? Well, times were hard enough for the poor in old Europe. Let him sweat more, but for whom? For himself of course, and good luck to him. Is there not plenty of Victoria land for every white man or black man that intends to grow his potatoes? Oh! leave the greens-growing to the well-disposed, to the well-affected, ye sturdy sons who pant after the yellow-boy. "Take your chance, out of a score of shicers, there is one 'dead on it,'" says old Mother Earth from the deep.

Sum total. — With the hard-working gold-digger, there is a solid barter possible. Hurrah! for the diggers.

The Argus persisting in "our own conceit," and misrepresenting, perverting, and slandering the cause of the diggers, ran foul, and went fast to leeward. Experience having instructed me at my own costs, that there cannot possibly exist much sympathy between flunkies and blueshirts, I can only guess at the compound materials hammered in the mortar of *The Argus* reporter on Ballaarat: —

lst. The land is the Queen's, and the inheritance of the Crown.

2nd. Who dares to teach the golden-lace the idea how to shoot?

3rd. Let learning, commerce, even manners die,

> But leave us our old nobility.

4th. *Sotto voce*: — In this colony, however, make money; honestly if possible, of course, but make money; or else the "vagabonds" here would humble down a gentleman to curry-powder diet.

5th. To put on a blue shirt, and rush in with the Eureka mob! fudge: "*odi profanum vulgus et arceo.*" There are millions of tons of gold dug out already, as much anyhow," as anyone can carry to Old England, and live as a lord, with an occasional trip to Paris and Naples, to make up for the time wasted in this colony.

Sum total. — Screw out of the diggers as much as circumstances will admit; they have plenty of money for getting drunk, and making beasts of themselves, the brutes!

To be serious; should a copy of this book be forgotten somewhere, and thereby be spared for the use of some southern Tacitus, let him bewail the perfidious mendacity of our times, whose characteristic is SLANDER, which proceeds from devil GROG; and the pair generate THE PROSPERITY OF THE WICKED. Here is a sample: —

On Saturday, September 29th, 1854, the members of the Local Court, Ballaarat, held a public meeting on the usual spot, Bakery-hill, for the purpose of taking the sense of their fellow miners, respecting the admittance or nonadmittance of the legal profession to advise or plead in said court. — See report in *The Star,* a new local paper, No. V, Tuesday, October 2nd.

Messrs. Ryce and Wall having addressed the meeting in their usual honest, matter-of-fact way: —

"Great Works" was shouted and immediately appeared C. Raffaello, member of the Local Court. He hoped, that if there were any Goodenough present that they would see and not mislay their notes while he briefly brought three things before the meeting; the first concerned the meeting and himself, the second concerned himself, and the third concerned those present.

The first was easily disposed of — have I, as I promised, done my duty as member of the Local Court to your satisfaction? (Yes, and cheers.) Very well, the second matter concerns myself — personally he was under no obligations to the lawyers — the services he received at the trial was done to him as a state prisoner, and not to Carboni Raffaello individually; when individually, he requested to be supplied with six pennyworth of snuff by Mr. Dunne, it was promised, but it never came to him. It would not have cost much to have supplied him, and it would have greatly obliged him, as habit had rendered snuff-taking necessary to him. With the permission of those present he would take a pinch now. (He took a pinch amidst laughter and cheers.)

The admission of lawyers into the Local Court would give rise to endless feuds, where valuable interests were concerned, and so much time would be lost in useless litigation. As he had no wish through any personal obligation to see the lawyers in the Local Court, and as he considered that it was for the advantage of the miners that they should not be admitted, he opposed their entrance.

The third matter concerned those present. What did they come to Australia for? Why, to improve their prospects in reality, though on shipboard they might say it was to get rid of the "governor," or to get clear of an ugly wife, and now that you are here are you to allow the Ballaarat lawyers to fleece you of your hard earnings? Not being fond of yabber-yabber he would simply ask: are you fairly represented by us? (Yes, yes.) If so then support us, and if we do not represent you we will resign. Don't say yes if you don't mean it, for I do not like yabber-yabber.

I beg to assert, that the above report is correct, as far as it goes. Some five hundred diggers were present. Now for the perversion from the reporter of *The Argus,* Melbourne, Tuesday, October 2.

"Carboni Raffaello, a foreigner [*a foreign anarchist, if you please, Mr. Editor*], then spoke in his usual style [*that is, sedition, revolution, and rebellion, that's it*], the principal (*sic*) points of his remarks being, that while incarcerated in the Melbourne gaol [*was it for common felony, or high treason?*] he was not supplied with snuff, though he had entreated his learned counsel, Mr. J. H. Dunne, for sixpenny worth. He [*Please, Raffaello or Dunne? fine pair together*] did not consider himself under any obligation to the lawyers: he [*but who? Dunne or Raffaello?*] was not fond of yabber-yabber."

Thus an honest man is brayed at by asses in this colony! The fun is odious and ridiculous enough.

When such reporters of the British press prostitute British ink, the only ink that dares to register black on white the name, word and deed of any tyrant through the whole face of the earth, and for the sake of a pair of Yankee boots, lower themselves to the level of a scribbler, thus affording to be audacious because anonymous, the British press in the southern hemisphere will be brought to shame, and Victoria cannot possibly derive any benefit from it.

Let the above observation stand good. I proceed with my work.

The Age was then just budding, and was considered, on the diggings the organ of the new chum Governor. *The Age* soon mustered a Roman courage in the cause of the diggers, and jumped the claims both of *The Herald* and *The Argus*; and though the "own correspondent," under the head of Ballaarat, be such a dry, soapy concern that will neither blubber nor blather, yet *The Age* remained the diggers' paper.

The Ballaarat Times was all the go, on the whole extent of the diggings. Soon enough the reporter, aye, the editor himself, will both appear in *propria persona*.

XVII.

ARCANE, IMPENETRABILI, PROFUNDE, SON LE VIE DI CHI DIE L'ESSER AL NIENTE.

WHEN our southern sky is overloaded with huge, thick, dark masses, and claps of thunder warn us of the pending storm, then a gale of wind is roaring in space, doing battle with the bush, cowing down man and beast, sweeping away all manner of rottenness. This fury spares not, and desolation is the threat of the thunder.

A kind Providence must be blessed even in the whirlwind. Big, big drops of rain fight their way through the gale; soon the drops muster in legions, and the stronger the storm, the stronger those legions. At last they conquer; then it pours down — that is, the flood is made up of legions of torrents.

Is the end of the world now at hand? Look at the victorious rainbow! it reminds man of the covenant of our God with Noah, not far from this southern land. The sun restores confidence that all is right again as before, and nature, refreshed and bolder, returns again to her work.

Hence, the storm is life.

Not so is the case with fire. Devouring everything, devouring itself, fire seems to leave off its frenzy, only to devour the sooner any mortal thing that comes in the way to retard destruction. A few embers, then a handful of ashes, are the sole evidence of what was once kingly or beggarly.

Fire may destroy, consume, devour, but has no power to reduce to "nothing."

Hence the calamity of fire is *death*.

The handful of ashes lie lifeless until a storm forces them into the living order of nature, which, when refreshed, has the power to ingraft those ashes to, and make them prosper with, the grain of mustard seed.

Hence death is life.

Such is the order of Providence. Now, good reader, watch the handful of ashes of what was once Bentley's Eureka Hotel.

XVIII.

PECUNIA OMNIA VINCIT.

IN the dead of the night after the burning of the Eureka Hotel, three men had been taken into custody, charged with riot, and subsequently committed to take their trial in Melbourne.

I think the diggers at this time seriously contemplated to burn down the Camp, and thus get rid in a blaze of all their grievances.

A committee for the defence of these men, met at the Star Hotel, and sent round to all the tents on Ballaarat for subscriptions. I contributed my mite, and then learned that VERN, KENNEDY, and HUMFFRAY were the triumvirate of said committee.

The following placard was posted throughout the gold-fields: —

"£500 REWARD
for the discovery, apprehension and conviction of the murderer
of James Scobie, found dead near the late Eureka Hotel,
etc., etc."

At one and at the same time, also, the following placards were posted at each prominent gum-tree on the goldfield: —

"£500 REWARD
increased by Government to

£1,600!!

for the apprehension and conviction of the robbers of
the Bank of Victoria."

A desperate deed was committed in broad mid-day; Monday, October 16th, in the Ballaarat township.

Four men in the garb of diggers, wearing sou'-wester hats, and having crape over their faces, entered the Bank of Victoria, and succeeded in carrying off property in notes and gold, to the amount of about £15,000.

Who would have told me then, that soon I should be messmate to those unknown audacious robbers, in the same gaol!!

Let's go to the public meeting in the next chapter.

XIX.

UNA SCINTILLA, SPARASI LA BOMBA,
SPALANCA A MULTITUDINI LA TOMBA.

THE following story was going the rounds of the Eureka. There was a licence-hunt; the servant of the Rev. P. Smyth, the priest of the Catholic church, Bakery-hill, went to a neighbouring tent to visit a sick man. While inside, a trooper comes galloping up at the tent-door, and shouts out, "Come out here, you d——d wretches! there's a good many like you on the diggings." The man came outside, and was asked if "he's got a licence?" The servant, who is a native of Armenia, answers, in imperfect English, that he is a servant to the priest. The trooper says, "Damn you and the priest;" and forthwith dismounts for the purpose of dragging Johannes M'Gregorius, the servant, along with him. The servant remonstrates by saying he is a disabled man, unable to walk over the diggings. This infuriates the trooper, he strikes and knocks down the poor disabled foreigner, drags him about, tears his shirt — in short, inflicting such injuries on the poor fellow, that all the diggers present cried out "shame! shame!"

Commissioner Johnson rides up, and says to the crowd about him, that he should not be interrupted in the execution of his "dooty." The priest hears of his servant's predicament, comes to the spot, hands a five-pound note to Johnson as bail for his servant's appearance the next day at the police-office.

The following morning, Johannes M'Gregorius is charged with being on the gold-fields without a licence. The poor foreigner tries to make a defence, but was fined five pounds. Commissioner Johnson now comes in and says, M'Gregorius is not charged with being without a licence, but with assaulting the trooper Lord — ridiculous! This alters the case. The trooper is called, and says the old story about the execution of "dooty," that is, licence-hunting.

A respectable witness takes his oath that he saw the trooper strike the foreigner with his clenched fist, and knock him down.

The end of the story is in the Ballaarat tune, then in vogue: "Fined £5; take him away."

XX.

PUBLIC MEETING.

Held at the Catholic Chapel, Bakery-hill, Wednesday, October 25th.

AFTER a good deal of pretty intelligible talk about the "helpless Armenian," the trooper Lord, and our respected priest; Thomas Kennedy, pouncing on the thing of the day proposed: —

"That it is the opinion of this meeting that the conduct of Mr. Commissioner Johnson towards the Rev. Mr. Smyth has been calculated to awaken the highest feeling of indignation on the part of his devoted flock: and to call upon the government to institute an inquiry into his (*gold-lace*) character, and to desire to have him at once removed from Ballaarat."

Carried unanimously.

The priest was requested to address the meeting.

FATHER PATRICIUS SMYTH, a native of Mayo, looks some thirty-five years old, and belongs to the unadulterated Irish caste — half-curled hair, not abundant, anxious semicircular forehead, keen and fiery eyes, altogether a lively interesting head. He is a Latin and Celtic scholar; and that excuses him for his moderate proficiency in modern languages. He was educated at Maynooth, the eye-sore of

Sabbatarians, and therefore believes it incontestable that the authority conferred on him by the Bishop must needs be derived from God; because the Bishop had been consecrated by the Pope, who — inasmuch as a second branch of the Prince of the Apostles never was heard of at the time of St. Augustin — is the successor of St. Peter, the corner stone on which OUR LORD did build the Christian church, and our Lord's warrant is written in St. John, chapter xiv, 24: *"Sermo quem auditis non est meus, sed ejus qui misit me, nempe Patris."* And so Father Smyth feels himself entitled to adopt what was said of the Divine Master, *"Docebat enim eos ut habens auctoritatem, non autem ut scribae."* St. Matthew, chap. vii, 29. Hence his preaching, though not remarkable for much eloquence, does not lull to sleep. There is no cant, and strange as it may appear, there is little argument in his short-framed sentences, because they are the decided opinion of his mind and the warm expression of his heart, anxious for the salvation of his flock, as he believes he will be called to account if any be lost. He, out of civility, may not object to hear what Paley or Butler has to say, but he scorns any conversation with Voltaire, and would see the fellow burnt, as in the times of old. His character was never impeached, because his conduct is an example to all of the strength of his faith. Either at the altar or at the table he forgets not that he belongs to the priesthood of Ireland, the "proved gold" of the Catholic church. His song is, *"Erin, my country,"* and *"I love thy green bowers,"* is the end of his story; which is a hint to me that this is not the place to say more for the peace of John Bull. Hence Ireland produced a Daniel O'Connell, but has not yet got the repeal.

Father Smyth, in addressing the meeting, spoke with coolness and forbearance, yet commendatory of the constitutional manner in which his congregation sought redress from the government, for the *insult* offered them, through his person, in the abuse of his servant by the trooper Lord. On concluding his address, he was warmly cheered, when the reverend gentleman and his friends adjourned to the parsonage, to partake of some refreshments

---

XXI.

PUBLIC MEETING.

Held on Bakery-hill, November 11th.

Political changes contemplated by
THE REFORM LEAGUE.

1. A full and fair representation. — *Dont you wish you may get it?*
2. Manhood suffrage. — *Thanks to the Eureka-boys, it costs now one pound. Cheap!*
3. No property qualification of members for the Legislative Council. — *The identical thing for "starring" on stumps to a fellow's heart's content.*
4. Payment of members. — *That's the accommodation!*
5. Short duration of Parliament. — *Increase the chances of accommodation, that's it.*

What was the freight per ton, of this sort of worn out twaddle imported from old England?

How much does this new chum's bosh fetch in the southern markets, and in the Victorian market particularly?

For my part I decline to answer, because I want to attend at the meeting. J. B. Humffray, is the Secretary of the League; his name is going now the round of the diggings; I wish to see the man in person; is he a great, grand, or big man? that's the question.

When you seen JOHN BASSON HUMFFRAY, you have at once before you a gentleman, born of a good old family; his manners confirm it, and his words indicate an honest benevolent heart, directed by a liberal mind, entangled perhaps by too much reading of all sorts, perplexed at the prosperity of the vicious, and the disappointment of the virtuous in this mysterious world of ours, but could never turn wicked, because he believes in the resurrection of life. He is looking some thirty five years old, his person is well proportioned, but inclining to John Bull's. His prepossessing countenance is made up of a fine forehead, denoting astuteness, not so much as shrewdness, how, when and whither to shift his pegs in the battle of life; of a pair of eyes which work the spell; of a Grecian nose; of a mouth remarkable for the elasticity of the lips, that make him a model in the pronunciation of the English language. His voice, that of a tenor, undulating and clear, never obstreperous, enables his tongue to work the intended charm, when his head puts that member into motion; but the semi-earnestness of his address, his cool sort of John Bull smile, betray that his heart does not go always with his head. Hence he has many enemies, and yet not one ever dared to substantiate a charge against his character; he has as many friends, but not *one friend*, because it is his policy ever to keep friendly, with redcoats and gold-lace, at one and the same time as with blueshirts and sou'-westers.

As I cannot possibly mean any thing dishonourable to our old mate, John Basson Humffray, I may here relate what his foes do say of him.

Suppose any given square and the four pegs to be:

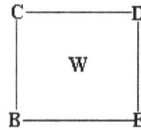

```
C————————D
|        |
|   W    |
|        |
B————————E
```

C., that is, the Camp; E., that is, the Eureka; D., that is, the doodledom of red-tape; and B., that is, blue-shirts.

Let W., that is work, be the central point at C, E, and D, B. Now: John is sinking at Eureka with the red cap; and Basson cracks some yabber-yabber at D, that is, getting a sip of Toorak small-beer, as aforesaid. Again: when Basson puts on a sou'-wester to go through the main-drift with blue-shirts, then John feels entitled to tramp up to Camp, and there, somewhere not far off, toast on the fourth of July a Doctor Kenworthy; soon after, however, said Johnny bends his way to shake hands with Signor Raffaello, at the old peg, Eureka, and helps him to rock the cradle. Further, to give evidence of his consistency, Humffray himself will express his sorrow to Peter Lalor for his loss of the left arm at same peg Eureka; and, to atone for past transgressions, he will soon after call in both the prodigal John and yabbering Basson, and with his whole heart and voice, strike up, "God Save the Queen," at peg Camp. As for bottoming his shaft at the central point Work, that's a different thing altogether; and yet it must be admitted that he is "all there" in his claim, *when the hole is bottomed*, especially if a drive is to be put in with his quill. Sum total: — He was, is, and ever will be, John Basson Humffray, Esquire, of Ballaarat; *Honi soi qui mal y pense,* because his friends want him in St. Patrick's Hall.

————

XXII.
STRIKE OFF A MEDAL IN COMMEMORATION.

WE are on Bakery-hill, though, attention. Immediate objects of the Reform League.

I. An immediate change in the management of the gold-fields, by disbanding the Commissioners (*undoubtedly the unanimous demand, or "desire" — if the word suit better the well-affected — of all blue-shirts*). Three cheers for Vern! Go it hearty! Fine fellow! Legs rather too long! never mind.

II. The total abolition of the diggers' and storekeeper's licence tax. (*Ah! Ah! prick John Bull at his £. s. d., that's the dodge to make him stir.*)

Three cheers for Humffray! Hurrah!

The whole of the grand talk of these Bakery reformers leagued together on its hill, can properly be framed in, on a "copper;" thus doing justice to all.

LET
a course of
action be decided on
and carried out unswervingly
until the heel of our oppressors
be removed from our necks.
DON'T LET THE THING DROP THROUGH,
for want of co-operation and support.
NOTA BENE.
2s. 6d. gentleman's ticket.
No admission for ladies at present.
Durum sed levius fit Patientia.
REMEMBER!
GOD HELPS HIM WHO HELPS HIMSELF (to the 2s. 6d.)
DO NOT LET
the word "British" become a bye-word.
AND ABOVE ALL LEAVE OFF SINGING
"Britons never, never shall be slaves,"
until you leave fondling
the chains which
prove the song
a lie,
a mockery,
a delusion,
a snare.

Great works!

XXIII.

ORTICA BALLAARATENSIS: PRIMA.

HERE is a plant of Cayenne pepper, growing in those days on Ballaarat: it withered some three months in limbo, but . . . oh yes, butt at it again.

Ballaarat Times, November 18, 1854.

"THE REFORM LEAGUE.

"There is something strange, and to the government of this country, something not quite comprehensible, in this League. For the first time in the southern hemisphere, a Reform League is to be inaugurated. There is something ominous in this; the word "League," in a time of such feverish excitement as the present, is big with immense purport (indeed!) Indeed, it would ill become the *Times* to mince in matter of such weighty importance. This League is not more or less that the germ of Australian independence (*sic*). The die is cast, and fate has stamped upon the movement its indelible signature. No power on earth can restrain the united might and headlong strides for freedom of the people of this country, and we are lost in amazement while contemplating the dazzling panorama of the Australian future (Great works). We salute the League [*but not the trio, Vern, Kennedy, Humffray*], and tender our hopes and prayers for its prosperity [*in the shape of a goodly pile of half-crowns*]. The League has undertaken a mighty task [*the trio'll shirk it though*], fit only for a great people — that of changing the dynasty of the country (*Great works*). The League does not exactly propose, nor adopt such a scheme, but we know what it means, the principles it would inculcate, and that eventually it will resolve itself into an Australian Congress." (*Great Works!!*)

𝔙𝔬𝔱𝔢 𝔣𝔬𝔯

HUMFFRAY to be Auctioneer,
KENNEDY to be Bellman,
VERN to be Runner,

of the "𝔖𝔱𝔞𝔯𝔯𝔦𝔫𝔤
𝔏𝔢𝔞𝔤𝔲𝔢."

XXIV.

ORTICA BALLAARATENSIS: SECUNDA.

OUT came the *Ballaarat Times*, Saturday, November 25, 1854. Work was stopped at every hole: the miners left the *deep* and mobbed together round any reader of the full report of the —

𝕿𝖗𝖎𝖆𝖑 𝖔𝖋
MR. AND MRS. BENTLEY,
Hanse, and Farrel,
FOR THE MURDER OF
JAMES SCOBIE.

Supreme Court, Melbourne.

GUILTY! of Manslaughter.
Mrs. Bentley scot-free.
His Honour considered their conduct was wanton and reckless. He should mark his sense of the outrage of which they have been found guilty, by passing on each of them a sentence of THREE (!) YEARS' IMPRISONMENT WITH HARD LABOUR ON THE ROADS.

𝕲𝖗𝖊𝖆𝖙 𝖂𝖔𝖗𝖐𝖘!

𝕿𝖗𝖎𝖆𝖑 𝖔𝖋
Fletcher, M'Intyre and Westerby,
for
BURNING THE EUREKA HOTEL.

Supreme Court, Melbourne.
Criminal Sittings.

GUILTY, with a recommendation to mercy!!
The Foreman of the Jury appended the following rider to the verdict: —
"The jury feel, in giving their verdict against the prisoners at the bar, that in all probability, they (the jury) should never have had that painful duty to perform, if those entrusted with the government offices at Ballaarat had done theirs properly."
His Honour said: THE SENTENCE of the Court is, that you, M'Intyre be confined in H. M. gaol, at Melbourne, for THREE MONTHS', *but I shall not subject you to labour*. (Great works!)

29

You, Fletcher, to four
months'; and you, Westerby,
to six months' confinement .
. . . . The
Executive was sufficiently
strong to punish those who
outrage the law! (Great
works at Toorak!)

La vita in grammatica,
Facil declinazione;
La vita poi in pratica,
Storta congiugazione:
Della vita lo spello dal mondo sciolto,
Al mondo vivi, poiché non sei sepolto.

———

XXV.

EPISTOLAM HANC MISI, TUNC BENE, NUNC VALDE AD OPUS.

"*Prepaid.*
"To W. H. ARCHER, Esq.
"Acting Registrar General,
"Melbourne."

" Ballaarat Gold-fields,
"Eureka, November 30, 1854.
"My dear Mr. Archer,
"I was in some anxiety about you; not receiving any answer to my letter of
the 17th October, and especially to that of the 22nd ditto. I was at Creswick's
Creek, when I was informed that Father Smyth had a letter for me, and last
Monday I returned to Ballaarat, where I received, through Messrs. Muir
Brothers, your letter of the 20th October. I am heartily glad to learn that you
are well, and now I suppose a few lines from me are as welcome to you as
ever.
"Somehow or other, verging towards the fortieth year of my age, having
witnessed strange scenes in this strange world, very, very different from my
dream of youth, I feel now more disposed to the sober reality of the things of
this life.
"However desponding and humiliating may be, as it really is, the sad
reflection, that at the enormous distance of *sixteen thousand miles* from dear

homes and dearer friends, people should be called upon to assemble, NOT to thank God Almighty for any special mercy, or rejoice over the first good harvest or vintage on this golden land; but melancholy is it to say, for the old purpose, as in olden times in the old country. "FOR THE REDRESS OF GRIEVANCES;" and so yesterday we had a monster meeting on Bakery-hill, and I was the delegate of upwards of one thousand foreigners, or 'aliens,' according to the superlative wisdom of your Legislative Council.

"The Camp was prepared to stand for the Colonial Secretary Foster! Yes; you may judge of the conduct of some officers sent to protect the Camp by the following: —

"On Tuesday Evening (November 28th), about eight o'clock, the Twelfth Regiment arrived from Melbourne. The expert cleverness of the officer in command, made the soldiers, riding in carts drawn by three horses each, cross the line exactly at the going-a-head end of the Eureka. An injudicious triumphant riding, that by God's mercy alone, was not turned into a vast funeral.

"From my tent, I soon heard the distant cries of 'Joe!' increasing in vehemence at each second. The poor soldiers were pelted with mud, stones, old stumps, and broken bottles. The hubbub was going on pretty desperate westward of the Hill and WE had hard work to preserve the peace; but at the upper end of the Hill, the game was going on upon a far more desperate scale. It appears that a party of Gravel-pits men had been in the bush for the purpose. They stopped a cart, pulled the soldiers out, robbed them of their ammunition and bayonets; in short, it was a hell of a row. All of us camping on the Hill were talking about this cowardly attack, when a detachment of said soldiers came up again, and the officer, a regular incapable, that is, a bully, with drawn sword began to swear at us, and called all of us a pack of scoundrels. He was, however, soon put to rights, by the whole of us then present offering ourselves to look out for the missing soldiers; and eventually, one of them was discovered in a deserted tent, another was found in a hole lower down the Warrenheip Gully, and so on. This disgraceful occurrence, coupled with the firing of guns and pistols, kept up the whole of the night, did not give us cheering hopes for the next day."

XXVI.

THE MONSTER MEETING.

Bakery-hill, Wednesday, November 29th.

(Letter continued.)

"All the diggings round about were deserted, and swelled the meeting, the greatest I ever witnessed in this Colony. At two o'clock there were about ten thousand men present! The Report of the Deputation appointed by the League to wait upon his Excellency, relative to the release of the three prisoners, M'Intyre, Fletcher, and Yorkie, was listened to with great anxiety."

GEORGE BLACK was the man of the day, and was received by the people with three hearty cheers.

From his outward appearance, one would take him for a parson, a Christian one, I mean; not a prebendary or a bishop. His English is elegant, and conscious of having received an education, and being born a gentleman, he never prostitutes his tongue to colonial phraseology. His reading must have been sober from his youth, for in conversation he indulges in neither cant nor romance; though, in addressing the people, he may use a touch of declamation stronger than argument. From the paleness of his cheeks, and the dryness of his lips, you might see that the spirit was indeed willing, though the flesh was weak. The clearness of his eyes, the sharpness of his nose, the liveliness of his forehead, lend to his countenance a decided expression of his belief in the resurrection of life. His principles are settled, not so much because that is required for the happiness of a good conscience, but because the old serpent has crammed the ways of man with so many deceits in this world of vanity and vexation of spirit, that a heart of the honesty of George Black, cannot possibly have any sympathy with the crooked ways of rogues and vagabonds; and so he is afflicted at their number and audacity, especially in this Colony. His disposition of mind makes him enthusiastic for the virtuous, his benevolent heart prevents him from proceeding to extremities with the vicious. Hence the *Diggers' Advocate*, of which he was the editor, though conducted with ability, failed, because he thought that gold-diggers interested themselves with true religion, as laid down in Saint James's Catholic Epistle; but he made a greater mistake in not taking into consideration that men, though digging for gold, do still pretend to some religious denomination or other. However, let him now address the Monster Meeting.

XXVII.

DIVIDE ET IMPERA.

(Letter continued,)

"Mr. Black explained the results of his mission by stating, that the Deputation was received by the Governor with much courtesy and urbanity, and that personally his Excellency had no objection to grant the public prayer. He further stated, that so far as he had an opportunity of judging of the Governor's disposition, his Excellency was in favour of the people, but that he was so surrounded by injudicious advisers, as to leave him entirely impotent in state matters. The great objection his Excellency seemed to entertain against the Deputation's claim, was what is termed want of courtesy in wording — for it must be understood that the Committee sent, not to petition and pray, but *demand* the release of the state prisoners; and the word *demand* was said to operate more against the Deputation than the very object of their mission. Upon hearing all these reasons, it was proposed to adopt the form of a memorial, and petition the Governor; but this proposition was furiously scouted, on the ground that it did not comport with the dignity of the League, first to demand, and afterwards to pray.

"Kennedy, along with the music of his rubbing the nails of the right hand against those of the left, blathered away in a masterly style for the benefit of the League.

"It was evident that there was a 'split' among the three Delegates; yet Mr. Humffray, who had been received by His Excellency, in an interview as a private digger, found favour among the assembly. J. B. Humffray plainly explained, and calmly made us understand, that Sir Charles was with us, and was determined to put an end to our grievances; and that he had appointed to this effect, a Commission of Inquiry, of popular men well known to us, and his Excellency had made up his mind to 'act accordingly.' The feverish excitement was subdued, and three hearty cheers were given for the New Chum Governor, amid the discharging of several guns and pistols."

I must here interrupt the meeting, drop the letter, and hereby assert: —

1st. Peter Lalor and myself, had never addressed any of the meetings, before this monster one.

2nd. Having made up my mind to return to Rome, the following Christmas, in accordance with my brother's desire; I had to attend to my work; hence, I had never taken any part in the agitation and to my knowledge, Peter Lalor neither.

3rd. I never was present at the Star Hotel and therefore, personally I know nothing of the boisterous Committee of the vaunting Reform League held there.

Corolarium. — I am not dead yet!

XXVIII.

L'UNION FAIT LA FORCE.

WE had better proceed with the meeting first, and with the letter afterwards. Peter Lalor proposed the following resolution: —

"That a meeting of the members of the Reform League be called at the Adelphi Theatre, on next Sunday, at 2 o'clock, to elect a Central Committee; and that each forty members have the power to elect one member for the Central Committee."

Being an old acquaintance of Peter, I supported the above resolution. He gave me his hand and pulled me up on the platform, from among the multitude. The whole of that Wednesday morning, my tent on the Eureka had been a regular Babel. Foreigners from all quarters of the globe and of the diggings, came to inquire from me what was the matter concerning so much excitement as then prevailed on Ballaarat. I translated for them the news from our *Ballaarat Times*, or from *The Geelong Advertiser's* clever correspondent. Thus, and thus alone, I became honourably their delegate, and subsequently interpreter to Lalor, the Commander-in-Chief; and I hereby express the hope that in time, Peter Lalor, though mutilated, may find at Toorak, a *little* more credit for his testimony than did that infernal spy, Goodenough. Anyhow, for the present, *Le Père Duprat*, a well known old hand, and respected French miner on Ballaarat, who was with me within the Eureka Stockade, and whose proposed plan for the defence, I interpreted to Lalor, is a living witness to the above. We must, however, attend to our Monster Meeting.

XXIX.

HEU MIHI! SERMO MEUS, VERITAS.

MY friends had requested me to come forward at the meeting, and here is my speech according to notes I had previously taken in my tent.

Gold-laced Webster, *I challenge contradiction*.

"I came from old Europe, 16,000 miles across two oceans, and I thought it a respectable distance from the hated Austrian rule. Why, then, this monster meeting to-day, at the antipodes? We wrote petitions, signed memorials, made remonstrances by dozens; no go: we are compelled to *demand*, and must prepare for the consequences.

"The old style: oppressors and oppressed. A sad reflection, very sad reflection, for any educated and honest man.

"For what did we come into this colony? *Chi sta bene non si move*, is an old Roman proverb. If then in old Europe, we had a bird in hand, what silly fools we were to venture across two oceans, and try to catch two jackasses in the bush of Australia!

"I had a dream, a happy dream. I dreamed that we had met here together to render thanks unto our Father in heaven for a plentiful harvest, such that for the first time in this, our adopted land, we had our own food for the year; and so each of us holding in our hands a tumbler of Victorian wine, you called on me for a song. My harp was tuned and in good order: cheerfully struck up,

'Oh, let us be happy together.'

Not so, Britons, not so! We must meet as in old Europe — old style — improved by far in the south — for the redress of grievances inflicted on us, not by crowned heads, but blockheads, aristocratical incapables, who never did a day's work in their life. I hate the oppressor, let him wear a red, blue, white, or black coat." — And here certainly, I tackled in right earnest with our silver and gold lace on Ballaarat, and called on all my fellow-diggers, irrespective of nationality, religion, and colour, to salute the '*Southern Cross*' as the refuge of all the oppressed from all countries on earth. — The applause was universal, and accordingly I received my full reward: Prison and Chains! Old style.

XXX.

THE REFORM LEAGUE, GRAPPLING WITH THE RIGHT "STARS."

MONSTER MEETING continued: —
Proposed and seconded by blather reformers; of course, Vern had his go: —

"That this meeting being convinced that the obnoxious licence-fee is an imposition and an unjustifiable tax on free labour, pledges itself to take immediate steps to abolish the same by at once burning all their licences; that in the event of any party being arrested for having no licence, that the united people will, under all circumstances, defend and protect them."

"That this meeting will not feel bound to protect any man after the 15th of December who shall not be a member of the Reform League by that day."

The Rev. Mr. Downing proposed as an amendment, that the licences should not be burned. Although the rev. gentleman was heard with patience and respect, a sullen excitement pervaded the whole assemblage while he spoke. Those even of his most devoted followers were of the opinion that his sentiments did not accord with the spirit of the times, and the result was that the rev. gentleman's amendment fell to the ground.

Here must not be forgotten a peculiar colonial habit. There was on the platform a sly-grog seller, who plied with the black-bottle all the folks there, and the day was very hot, the sun was almost burning.

XXXI.

SI CESSI IL PIANTO, L'IRA SI GUSTI.
LO SCHIAVO CHE VUOL FINIR LE SUE PENE,
VENDETTA GRIDANDO AL DIO DE GIUSTI,
DEVE SCHIANTAR LE PROPRIE CATENE.
CUORE! SI VADA, VEDASI, SI VINCA *(bis.)*

IN spite of all that, however, Timothy Hayes, the chairman — who by-the-bye, discharged the duties of the chair in that vast assemblage, with ability and tact, spoke like a man, as follows: —

"Gentlemen, many a time I have seen large public meetings pass resolutions with as much earnestness and unanimity as you show this day; and yet, when the time came to test the sincerity, and prove the determination necessary for carrying out those resolutions, it was found then that 'the spirit, indeed, is willing, but the flesh is weak.' Now, then, before I put this resolution from the chair, let me point out to you the responsibility it will lay upon you (hear, hear). And so I feel bound to ask you, gentlemen, to speak out your mind. Should any member of the League be dragged to the lock-up for not having the licence, will a thousand of you volunteer to liberate the man?"

"Yes! Yes!"

"Will two thousand of you come forward?"

"Yes! Yes! Yes!"

"Will four thousand of you volunteer to march up to the Camp, and open the lock-up to liberate the man?"

"Yes! yes!" (the clamour was really deafening.)

"Are you ready to die?" shouted out our worthy chairman, stretching forth his right hand, clenched all the while; "Are you ready to die?"

"Yes, Yes! Hurrah!"

This general decided clamour put our Tim in such good spirits, that, in spite of the heat of the sun and the excitement of the day, he launched in the realm of crowned poets, and bawled as loud as if he wanted the head-butler at Toorak to take him a quart-pot of small-beer —

> "On to the field, our doom is sealed,
>   To conquer or be slaves;
> The sun shall see our country free.
>   Or set upon our graves."

(Great works!)

No one who was not present at that monster meeting, or never saw any Chartist meeting in Copenhagen-fields, London, can possibly form an idea of the enthusiasm of the miners of Ballaarat on that 29th of November. A regular volley of revolvers and other pistols now took place, and a good blazing up of gold-licences. When the original resolutions had all been passed, Mr. Humffray moved a vote of thanks to Mr. Ireland, for his free advocacy of the state prisoners. The meeting then dissolved, many of them having previously burned their licences, and thus virtually pledging themselves to the resolution adopted, which might be said to have been the business of the day. Nothing could exceed the order and regularity with which the people, some fifteen thousand in number, retired.

XXXII.

ECCO TRONCATO IL CANTO
PER RITORNARE AL PIANTO.

MY letter to Mr. Archer continued: —

"Thanks be to God, the day passed 'unstained,' a glorious day for Victoria when the SOUTHERN CROSS was first unfolded on Ballaarat; gathering round itself all the oppressed of the world.

"The whole purpose of the meeting was, that a Reform League be formed and fully organised to carry out the clearance of all our grievances, on the old style of the Corn Law League in Great Britain.

"Next Sunday, we leaguers — *(I took out a ticket of membership from Reynolds, one of the treasurers, and paid my 2s. 6d. on that very day, November 29th, precisely, on the platform of the meeting)* — have a meeting at two o'clock at the Adelphi to organise the people and appoint a responsible executive committee. I am the old delegate to it, and therefore I shall be able to give you, Mr. Archer, a full answer to your letter of the 24th instant."

Mark this, good reader!

"1. Meanwhile, privately, as an old Ballaarat hand, I beg respectfully to convey to you, to employ your influence and reach the ears of the Lieutenant-Governor. The licence-fee, as a tax, is perhaps a cause of growling like any other tax in Great Britain or elsewhere in the world; but, on the gold-fields, has become an 'abomination.' The inconvenience in, the Camp-insolence at, our getting it, the annoyance and bore for showing it, when asked by some 'pup' of a trap whilst at our work; the imbecility and arrogance of so many commissioners and troopers uselessly employed for the purpose, etc., etc.; make the gold-licence an abomination to the honest digger. The Vandemonian, you know, never dreamt of taking out a licence, of course not.

"Paramount is this grand consideration: John Bull, rather of a doggish nature, will growl to himself if left alone picking his bone: the passport system is a bone that he will not pick; no, no ways and under no shape whatever — *I know it by experience.*

"2. A memorial to his Excellency for the release of the three prisoners under sentence for burning the Eureka Hotel, is, through Humffray, in course of signature. It is our earnest desire that his Excellency may show mercy; though it may appear, that he would do thus an act of justice to the diggers, considering how rightly they guessed the Bentley affair.

"3. The whole pack, commissioners, troopers and traps on the Ballaarat Camp, with the exception of magistrate Hackett, are detested by the diggers: there will be eternal discontent as long as Rede and fraternity are lodging over that way. The whole Camp had better be changed at once, and entrusted to good experienced hands and honest men. Perhaps Sir Charles may turn into a Diogenes in vain — *'nil desperandum.'* There are now and then honest men to be found even in this colony. "

Good reader, listen to me: I shall tell you no lie: do not lose sight of the above letter: I intend to give the end in the next chapter: meanwhile, fill the pipe, let's have a "blow" together.

———

XXXIII.

MISTERO! S'APRE MENDACIA, VIOLENTE
STRADA MAESTRA IN CITTA E CAMPAGNA
LA VERITA, SE DOCILE, QUADAGNA
A PASSO LO STRADELLO LENTAMENTE.

(Translated in the text of my first chapter.)

ON Thursday morning, November 30th, at sunrise, I was at my work, as usual.
I assert, as an eye-witness, that most of the hands on the Eureka came to their work, and worked as usual.

Whilst having a "blow," we would talk over again about the monster meeting of yesterday, thus spinning a yarn in the usual colonial style.

The general impression was, that as soon as government knew in Melbourne the real state of the excited feelings of the diggers, the licence-hunt would be put a stop to.

Towards ten o'clock was my hour for a working-man's breakfast. I used to retire to my tent from the heat of the mid-day, and on that same Thursday I set about, at once, to end my letter to Mr. Archer, because I was anxious to forward it immediately to Melbourne.

Good reader, I copy *now*, word for word, the scrawl *then* penned, in great haste and excitement.

"Thursday, November 30th, 1854.

"Just on my preparing to go and post this letter, we are worried by the usual Irish cry, to run to Gravel-pits. The traps are out for licences, and playing hell with the diggers. If that be the case, I am not inclined to give half-a-crown for the whole fixtures at the Camp.

"I must go and see 'what's up.'
Always your affectionate,
(*Signed*) "CARBONI RAFFAELLO."
"W. H. ARCHER, Esq.,
"Acting Registrar-General, Melbourne."

WHY this identical letter of mine — now in the hands of James Macpherson Grant, M.L.C., Solicitor, Collins-street, where it will remain till Christmas for inspection, to be then returned to the owner — *was not produced at my STATE TRIAL, was, and is still, a MYSTERY to me!*
Let's run to Bakery-hill.

_____

XXXIV.

QUOS VULT PERDERE DEUS DEMENTAT.

WHAT'S up? a licence hunt; old game. What's to be done? Peter Lalor was on the stump, his rifle in his hand, calling on volunteers to "fall in" into ranks as fast as they rushed to Bakery-hill, from all quarters, with arms in their hands, just fetched from their tents. Alfred, George Black's brother, was taking down in a book the names of divisions in course of formation, and of their captains.

I went up to Lalor, and the moment he saw me, he took me by the hand saying, "I want you, Signore: tell these gentlemen, *pointing to old acquaintances of ours, who were foreigners;* that, if they cannot provide themselves with fire-arms, let each of them procure a piece of steel, five or six inches long, attached to a pole, and that will pierce the tyrants' hearts." Peter of course spoke thus in his friendly way as usual towards me. He was in earnest though. The few words of French he knows, he can pronounce them tolerably well, but Peter is no scholar in modern languages; therefore he then appointed me his aide-de-camp, or better to say his interpreter, and *now I am proud to be his historian.*

Very soon after this, all the diggers "fell in" in file of two-a-breast, and marched to the Eureka.

CAPTAIN Ross of Toronto, was our standard-bearer. He hoisted down the Southern Cross from the flag-staff and headed the march.

Patrick Curtain, the chosen captain of the pikemen, gave me his iron pike, and took my sword to head his division; I "fell in" with John Manning who also had a pike, and all of us marched in order to the Eureka.

I assert as an eye witness, that we were within one thousand in the rank with all sort of arms, down to the pick and shovel.

We turned by the Catholic church, and went across the gully. Of this I have perfect recollection: when the "Southern Cross" reached the road, leading to the Eureka on the opposite hill, the file of two-a-breast crossing the gully, extended backwards up to the hill where the Catholic church stands. I took notice of the circumstance at the time.

We reached the hill where was my tent. How little did we know that some of the best among us had reached the place of their grave! Lalor gave the proper orders to defend ourselves among the holes in case the hunt should be attempted in our quarters.

The red-tape was by far too cunning this time; red-coats, traps and troopers had retired to the Ballaarat Camp, and wanted a "spell."

We determined, however, to put an end to their accursed licence-hunting, mock riot-act chopping, Vandemonian shooting down our mates in Gravel-pits.

———

XXXV.

AD OPUS CONCILIUM STATUTUM.

PETER LALOR, at our request, called in all the captains of division, then present, and the chief persons who had taken part in the movement. We entered a room some twelve feet square, in *Diamond's* store. An old European fox for such occasions, I took the right sort of precautions, that no spy might creep in among us. Black bottles and tumblers were placed on the table, as a blind to any intruder; "*et nunc satis, profani vulgus causa,*" we proceeded to business.

Present —

1. There was one, whom it is not prudent to mention just now,

2. Near him was a thick, short-necked, burly individual; his *phisiog* indicated at once that he was a priest-ridden. I won't trouble myself about his name.

3. I'll begin with TIMOTHY HAYES. He was born in Ireland, but his outward appearance is that of a noble fellow — tall, stout, healthy-looking man, giving himself the airs of a high-born gentleman, fit to rule, direct, superintend, not to work; that's quite another thing. Of a liberal mind, however, and, above all, of a kind heart, and that covers a multitude of sins.

4. EDWARD THONEN, a native of Elbertfeld, Prussia, five feet high, some thirty years old, thin, but robust, of vigorous health, used no razor. His eyes spoke determination and independence of character. One day in November, 1853, he called with his lemonade kegs at my hole in Sailor's Gully. A mate was served with a glass of lemonade — halloo! he must help at the windlass just at the moment he was tendering payment, and the shilling fell to the ground. Some words passed to the effect that six-pence a glass should be enough for lemonade. Thonen asked for his shilling; my mate directed him where the shilling lay; Thonen would see him d———d first before picking up his money like a dustman, and went away. I sent that identical shilling (stamped 1844), along with my little gold, to Rome; most astonishing! I had the presentiment at the time that I should have had occasion to relate the story. There was no mate on the gold-fields to match Thonen at chess-playing. He would turn his head, allow his opponent the move, and then he would give such a glance on the chess board, that the right piece would jump to the right place, as it were of its own accord. Shrewd, yet honest; benevolent, but scorning the knave; of deep thought, though prompt in action; Thonen possessed the head belonging to that cast of men whose word is their bond.

5. JOHN MANNING, born in Ireland, and an Irishman to the back-bone, appeared above forty years of age. His head was bald, perhaps from thinking three times more than he ought; his forehead showed intelligence, but care was there with the plough — the plough of dreaming too much of virtue, believing the knaves are *not* the majority on earth. He had come young to this colony, had passed hard days, and so he had got the colonial habit, now and then, "*Divo jucundo Baccho cultum prestare;*" hence his hair was fast turning grey. He was a self-educated man, but

wanted judgment to discipline his fermenting brain, for the control of his heart, which was good, honest, always warm, affectionate to man, woman, and child. When he took his quill he was "all there," but soon manifested the sort of reading of his youth; and experience, however hard, had not yet taught him the sober reality of the things of the world — that is, he had remained an Irishman, not John Bullised.

6. Oh! you long-legged VERN! with the eyes of an opossum, a common nose, healthy-looking cheeks, not very small mouth, no beard, long neck for Jack Ketch, broad shoulders, never broken down by too much work, splendid chest, long arms — the whole of your appearance makes you a lion amongst the fair sex, in spite of your bad English, worse German, abominable French. They say you come from Hanover, but your friends have seen too much in you of the Mexico-Peruvian. You belong to the school of the "Illuminated Cosmopolitans;" you have not a dishonest heart, but you believe in nothing except the gratification of your silly vanity, or ambition, as you call it.

7. The next was a skinny bouncing curl who affected the tone and manners of a Californian; he acted throughout the part of a coward, I scorn to mention his name.

8. Thank God there is among us a man; not so tall as thick, of a strong frame, some thirty five years old, honest countenance, sober forehead, penetrating look, fine dark whiskers. His mouth and complexion denote the Irish, and he is the earnest, well-meaning, no-two-ways, non-John-Bullised Irishman, PETER LALOR, in whose eyes, the gaseous heroism of demagogues, or the knavery of peg-shifters is an abomination, because his height of impudence consisted in giving the diggers his hand, and leaving with them his arm in pawn, for to jump the Ballaarat claim in St. Patrick's Hall. More power to you Peter! Old chummy, smother the knaves! they breed too fast in this colony.

9. Myself, CARBONI RAFFAELLO, DA ROMA; Member of the College of Preceptors (1850), Bloomsbury-square, professor, interpreter and translator of the Italian, French, Spanish and German Language into English or vice versa late of 4, Castle-court, Birchin-lane, Cornhill, London; now, gold-digger of Ballaarat, was present.

10. PATRICK CURTAIN, an old digger, well known among us; at the time a storekeeper; husband and father of a beloved family. His caste is that of the Irishman — John-bull; tall, robust, some forty years old; he is no friend to much yabber-yabber; of deep thinking, though very few can guess what he is thinking of. He smiles but never laughs to his heart's content. Curtain was captain, and subsequently lieutenant of the pikemen division, when they chose HANRAHAN for their captain. Said pikemen division was among the first that took up arms on Thursday, November 30th, immediately after the licence-hunt. It was formed on Bakery-hill, and received Lalor on the stump with acclamation. It increased hourly and permanently; was the strongest division in the Eureka stockade; in comparison to others, it stood the most true to the "Southern Cross," and consequently suffered the greatest loss on the morning of the massacre. Now, to explain how both its gallant leaders escaped unhurt, safe as the Bank, so that a few weeks afterwards, both were working happy and jolly in broad day-light on Gravel-pits, within a rifle shot from the Camp, that would be a job of a quite different kind just

at present: sufficient the trouble to mention; that when I came out of gaol, I met them both in a remunerative hole in Gravel-pits, as aforesaid.

11. 12. There were two other individuals of the John-bull caste, perhaps cross-breed, who had taken up arms in the cause of the diggers, because their sly-trade was flagging; but, as a rotten case abides no handling, I will let them pass.

Manning, handed over to Lalor the motion drawn up in my tent. Here it is: —

Proposed by John Manning,

Seconded by Carboni Raffaello,

"I. That Peter Lalor has acted worthy of the miners of Ballaarat, in organising the armed men on Bakery-hill, against the wanton aggression from the Camp this morning.

"II. That he be desired to call in all captains of division now present on the spot, as well as other persons of importance, well-known good-wishers to the cause of the diggers.

"III. That said parties constitute the council-of-war for the defence.

"IV. Lalor to be the president pro. tem.

"V. That he proceed at once to the election of the Commander-in-Chief, by the majority of votes."

Lalor tore up immediately the slip of paper containing the above motion, because he did not think it prudent to leave written things about in a public store. I transcribe it from the scrap left among the papers in my tent.

XXXVI.

QUOUSQUE TANDEM ABUTERE, TOORAK, PATIENTIA NOSTRA?

LALOR rose, and said: —

"Gentlemen, I find myself in the responsible position I now occupy, for this reason. The diggers, outraged at the unaccountable conduct of the Camp officials in such a wicked licence-hunt at the point of the bayonet, as the one of this morning, took it as an insult to their manhood, and a challenge to the determination come to at the monster meeting of yesterday. The diggers rushed to their tents for arms, and crowded on Bakery-hill. They wanted a leader. No one came forward, and confusion was the consequence. I mounted the stump, where you saw me, and called on the people to 'fall in' into divisions, according to the fire-arms they had got, and to chose their own captains out of the best men they had among themselves. My call was answered with unanimous acclamation, and complied to with willing obedience. The result is, that I have been able to bring about that order, without which it would be folly to face the pending struggle like men. I make no pretensions to military knowledge. I have not the presumption to assume the chief command, no more than any other man who means well in the cause of the diggers. I shall be glad to see the best among us take the lead. *In fact, gentlemen, I expected some one who is really well known* (J. B. Humffray?) *to come forward and direct our movement!* However, if you appoint me your commander-in-chief, I shall not shrink; I mean to do my duty as a man. I tell you, gentlemen, *if once I pledge my hand to the diggers, I will neither defile it with treachery, nor render it contemptible by cowardice.*"

Brave Peter, you gave us your hand on the Eureka, and left there your arm: an incontestible evidence of Lalor's Pledge.

Manning then proposed Raffaello, and pointed at his scars as an evidence of his tiger-pluck against the hated Austrian rule, which was now attempted, in defiance of God and man, to be transplanted into this colony.

I declined, because, during the past winter, I had over tasked my physical strength, and did not possess that vigour essential to such an emergency. Confidence is the bond necessary between the soldier and his officer. It was my decided opinion, however much a foreigner may be respected on the gold-fields, that the right man should be taken from among Britons.

Vern here began a portentous lecture on military science, military discipline, military tactics, and other sorts of militaryism, but his English was so wretched, his ideas so sky-blathering, his martial ardour so knocking down, that no one could make anything out of his blabberdom.

Of this I have perfect recollection. He was boasting eternally of *his* German rifle-brigade! 500 strong. That he had this brigade he urgently asserted; but where it was, that's the rub!

No possible inquiry from Lalor could get at the bottom of Vern's prodigal brigade. Is, then, the grand secret buried within Vern's splendid chest? No; I mean to reveal it at four o'clock, Saturday, December 2nd.

Carboni Raffaello, who had heard heaps of cant in old Europe, did count for nothing the oceanic military knowledge of Vern, in spite of his big trail-sword, that made more jingling than enough.

I commended, in high terms, the conduct of Lalor during the morning, and it was my impression that he possessed the confidence of the diggers and should be their Commander-in-chief.

Thonen seconded the motion. The first "unnamed," shewed approbation, and the appointment was carried by a majority of eleven to one.

Peter Lalor thanked the council for the honour conferred on him, assured the members that he was determined to prepare the diggers to resist force by force.

It was perfectly understood, and openly declared, in this first council-of-war, that we meant to organise for *defence*, and that we had taken up arms for no other purpose.

The council adjourned to five o'clock in the evening.

———

XXXVII.

LALOR STUMP, BAKERY-HILL.

Brave LALOR —
    Was found "all there,"
    With dauntless dare,
    His men inspiring;
      To wolf or bear,
      Defiance bidding,
      He made us swear,
    Be faithful to the Standard,
    For Victory or Death! (*bis* last 2 lines)

ON that Thursday, November 30th, more memorable than the disgraced Sunday, December 3rd, the SUN was on its way towards the west: in vain some scattered clouds would hamper its splendour — the god in the firmament generously ornamented them with golden fringes, and thus patches of blue sky far off were allowed to the sight, through the gilded openings among the clouds.

The "SOUTHERN CROSS" was hoisted up the flagstaff — a very splendid pole, eighty feet in length, and straight as an arrow. This maiden appearance of our standard, in the midst of armed men, sturdy, self-overworking gold-diggers of all languages and colours, was a fascinating object to behold. There is no flag in old Europe half so beautiful as the "Southern Cross" of the Ballaarat miners, first hoisted on the old spot, Bakery-hill. The flag is silk, blue ground, with a large silver cross, similar to the one in our southern firmament; no device or arms, but all exceedingly chaste and natural.

CAPTAIN ROSS, of Toronto, was the bridegroom of our flag, and sword in hand, he had posted himself at the foot of the flag-staff, surrounded by his rifle division.

PETER LALOR, our Commander-in-chief, was on the stump, holding with his left hand the muzzle of his rifle, whose but-end rested on his foot. A gesture of his right hand, signified what he meant when he said, "It is my duty now to swear you in, and to take with you the oath to be faithful to the Southern Cross. Hear me with attention. The man who, after this solemn oath does not stand by our standard, is a coward in heart.

"I order all persons who do not intend to take the oath, to leave the meeting at once.

"Let all divisions under arms 'fall in' in their order round the flag-staff."

The movement was made accordingly. Some five hundred armed diggers advanced in real sober earnestness, the captains of each division making the military salute to Lalor, who now knelt down, the head uncovered, and with the right hand pointing to the standard exclaimed a firm measured tone: —

"WE SWEAR BY THE SOUTHERN CROSS TO STAND TRULY BY EACH OTHER, AND FIGHT TO DEFEND OUR RIGHTS AND LIBERTIES."

An universal well rounded AMEN, was the determined reply; some five hundred right hands stretched towards our flag.

The earnestness of so many faces of all kinds of shape and colour; the motley heads of all sorts of size and hair; the shagginess of so many beards of all lengths and thicknesses; the vividness of double the number of eyes electrified by the magnetism of the southern cross; was one of those grand sights, such as are recorded only in the history of "the Crusaders in Palestine."

XXXVIII.

UN BON CALCIO, E LA CANAGLIA, STRONCA VA COME LA PAGLIA.

THE drill afterwards was gone through with eagerness.

Another scene, though of a different kind all together, was going on at a corner of the above picture.

Judas Iscariot, "Goodenough," was among us, in the garb of a fossiker; he appeared to me, then, to be under the influence of drink; so Vandemonian-like were his shouts about standing up and fighting for rights and liberties; and burning down the camp in a blaze like the late Eureka Hotel.

Mind good reader, I tell you no joke, I am not in humour just now to spin a yarn. — I wished to shame the fellow for his villany on such a solemn occasion. The fellow persisted in his drunken shouts. I lost my temper, and gave the scoundrel such a respectable kick, in a less respectable region, with a most respectable boot of mine, that it served me right when both my new watertight boots were robbed from my shins by Goodenough's satellites on the subsequent Sunday, at the Ballaarat Camp.

The Thursday's sun is setting: we returned to the Eureka. I had to attend the "Council for the Defence."

XXXIX.

DISCIPLINA, SUPREMA LEX IN BELLO.

IN the afternoon, our camp on the Eureka was enclosed in by slabs, near-handy thrown down at random. All diggers who had been able to procure fire-arms kept coming in, in right earnest, and formed new divisions. The pikemen grew stronger and stronger. Drilling was tolerably progressing. We were of all nations and colours. Lalor gave me his consent and order to direct all foreigners in their respective language, however little they knew of the English, to fall in in divisions according to

45

the arms they had got; and here I solemnly declare, to whomsoever it may concern, that up to four o'clock of Saturday there was not one single division distinguished by nationality or religion.

The armed men numbered now (six o'clock) about five hundred.

Vern's gall was fermenting, but on PETER LALOR being proclaimed Commander-in-chief, the appointment was ratified by hurrah! from the diggers.

There was such a decided intention to do "something" with the strong arm, and at once, that I was called on the stump. I requested the diggers to give us time for deliberation, and pledged my word that I would inform them of the result. "Go a-head! Great works!" was the shout.

XL.

BEATI QUI SUNT PACIFICI, QUONIAM FILII DEI VOCABUNTUR.

IT was dark: myself took the light in the council room.

Father P. Smyth and Mr. George Black were present; both looking serious and anxious.

All of us were much concerned, and felt the responsibility of our position. By this time the diggers from all parts had swelled to the number of eight hundred. They were not clamorous, they wanted to know what was determined on by the leaders.

Proposed by Black, seconded by Manning, "That a deputation from the armed diggers, should be forthwith sent to the Camp —

"1. To demand — that was our temper in those days — the immediate release of those diggers who had been dragged to the lock-up in the morning hunt, for want of the licence.

"2. To demand from Commissioner Rede a pledge not to come out any more for licence-hunting."

Two of us were to form the deputation, and proceed at once.

Father Smyth proposed Mr. Black, LALOR proposed Signor Raffaello: agreed to unanimously. This news, being made public to the diggers, was well received by all; and the council kept sitting until our return.

The deputation was accompanied by Father Smyth. It was a starry night, and rather cold; the moon shone in all its southern splendour. On approaching the main road, the noisy band of Row's Circus, and the colonial cursing and shouting from inveterate grog-bibbers, forced into my mind the meditation, "*Unde bella et pugna infer vos?*" etc. — James, chap. iv.

We met here and there several groups, who were anxiously discussing the events of the day, and the probable consequences. Mr. Black kindly and plainly informed them of our mission. On reaching the bridge, we found it guarded by the police. Father Smyth had an easy pass, and went by himself to speak first at head-quarters, for the safety of our persons.

XLI.

THE EUREKA STOCKADE.

THE CONSEQUENCE OF SOME PIRATES WANTING ON QUARTER-DECK A REBELLION.

AT last the deputation was before King Rede, whose shadow by moonshine, as he held his arm *a la Napoleon*, actually inspired me with reverence; but behold! only a *marionette* was before us. Each of his words, each of his movements, was the vibration of the telegraphic wires directed from Toorak. He had not a wicked heart; some knew him for his benevolence, and he helped many an honest digger out of trouble. Once I had seen him with my mate, Paul Brentani, about manufacturing bricks from the splendid clay of Gravel-pits. Mr. Rede received us as a gentleman, and, by way of encouragement, said to Paul, *"Je veux bien vous aider, car tout est encore a batir a Ballaarat, et il nous faut des briques — revenez me voir."* And yet, on the gold-field, this man was feared by the few who could not help it, respected by the many — detested by all, because he was the Resident Commissioner — that is, all the iniquities of officialdom at the time were indiscriminately visited on his gold-lace cap, which fact so infatuated his otherwise not ordinary brains, that they protruded through his eyes, whenever he was sure he had to perform a *dooty*. I would willingly turn burglar to get hold of the whole of the correspondence between him and Toorak. I feel satisfied I would therein unravel the MYSTERY of the Eureka massacre.

Rede, after all, was neither the right man, nor in the right place, for Sir Charles Hotham.

Sub-inspector Taylor, with his silver-lace cap, blue frock, and jingling sword, so precise in his movement, so Frenchman-like in his manners, such a puss-in-boots, after introducing the deputation, placed himself at the right of the Commissioner, and never spoke; though, on accompanying us from the bridge, having recognised me, he said, "We have been always on good terms with the diggers, and I hope we may keep friends still;" — and gave a smile of sincerity. He, perhaps, was ignorant, as well as the deputation, that, on quarter-deck, some pirates wanted a rebellion.

At the left of Mr. Rede, there was a gentleman who inspired us with confidence. His amiable countenance is of the cast that commands respect, not fear. The ugliness of his eyes prejudices you against him at first; let him, however, turn them upon you in his own benevolent way, you are sure they mean no harm: within a pair of splendid whiskers, of the finest blond, there is such a genteel nose and mouth, such a fine semi-serious forehead, that the whole is the expression of his good sound heart, that loves truth, even from devils. It was CHARLES HENRY HACKETT, police magistrate.

The place of our palaver was exactly one yard down hill, near the old gum tree, in front of the present Local Court.

Mr. Rede asked our names, and cautioned us that our message would be reported at head-quarters. He who had a gang of the vilest spies at his bidding, perhaps, indeed, forced upon him, now suspected us as such, and told us pretty plainly, that he thought it *not* prudent to take us to his residence, the camp being prepared against a supposed attack from the diggers.

XLII.

INVANUM LABORAVIMUS.

MR. BLACK began, in plain and straightforward language, to make a truthful statement of the exasperated feelings of the diggers, courageously censuring the conduct of the Commissioner in his licence-hunt of the morning, reminding him of the determination with which the diggers had passed the resolutions at the monster meeting of yesterday. "To say the least, it was very imprudent of you, Mr. Rede, to challenge the diggers at the point of the bayonet. Englishmen will not put up with your shooting down any of our mates, because he has not got a licence."

Mr. Rede: "Now Mr. Black, how can you say that I ever gave such an order as to shoot down any digger for his not having a licence?" and he proceeded to give his version of the occurrence. Master Johnson wanted a little play, and rode licence-hunting; was met with impertinent shouts of "Joe, Joe," and reported a riot. Daddy Rede must share in the favourite game, and rode to crack the riot act. The red-coats turned out. The diggers mobbed together among the holes, and several shots were fired at the traps. The conclusion: Three of the ring-leaders of the mob had been pounced upon, and were safe in chokey.

Mr. Black manfully vindicated the diggers, by observing how they had been insulted; that Britons hated to be bullied by the soldiery, and concluded by stating our first "demand."

Mr. Rede, startled at our presumption, breathed out "Demand! — First of all, I object to the word, because, myself, I am only responsible to government, and must obey them only: and secondly, were those men taken prisoners because they had not licences? Not at all. This is the way in which the honest among the diggers are misled. Any bad character gets up a false report: it soon finds its way in certain newspapers, and the Camp officials are held up as the cause of all the mischief."

Mr. Black would not swallow such a perfidious insinuation.

Mr. Rede continued: "Now, Mr. Black, look at the case how it really stands. Those men are charged with rioting; they will be brought before the magistrate, and it is out of my power to interfere with the course of justice."

Mr. Hackett spoke his approbation to the Commissioner.

Mr. Black: "Will you accept bail for them to any amount you please to mention?"

A consultation ensued between Rede and Hackett. Bail would be accepted for two of the prisoners. Father Smyth would bring the required sureties to-morrow morning.

Mr. Black proceded to our second demand.

Mr. Rede took that for a full stop; and launched into declamation: "What do you think, gentlemen, Sir Charles Hotham would say to me, if I were to give such a pledge? Why Sir Charles Hotham would have at once to appoint another Resident Commissioner in my place!" and concluded with the eternal cant of all silver and gold lace, "I have a *dooty* to perform, I know my duty, I must *nolens volens* adhere to it."

In vain Mr. Black entered the pathetic; and expostulated with the Commissioner, who had it in his power to prevent bloodshed.

Mr. Rede: "It is all nonsense to make me believe that the present agitation is intended solely to abolish the licence. Do you really wish to make me believe that the diggers of Ballaarat won't pay any longer two pounds for three months? The licence is a mere cloak to cover a democratic revolution."

Mr. Black acknowledged, that the licence fee, and especially the disreputable mode of collecting it at the point of the bayonet, were *not* the only grievances the diggers complained of. They wanted to be represented in the Legislative Council; they wanted to "unlock the lands."

Carboni Raffaello, who had yet not opened his mouth, said: "Mr. Rede, I beg you would allow me to state, that the immediate object of the diggers taking up arms, was to resist any further licence-hunting. I speak for the foreign diggers whom I here represent. We object to the Austrian rule under the British flag. If you would pledge yourself not to come out any more for the licence, until you have communicated with *Son Excellence*, I would give you my pledge . . . — (I meant to say, that I was willing to pledge myself, and try if possible to assuage the excitement, and wait till 'our Charley' had sent up a decided answer") — but I was instantly interrupted by Father Smyth who addressed me imperatively: "Give no pledge: sir, you have no power to do so."

This interruption, which I perfectly recollect, and the circumstance that on our going and returning, the said Father Smyth continually kept on a *sotto voce* conversation with Mr. Black only, were, and are still, *mysteries* to me.

Mr. Rede, who had not failed to remark the abruptness with which Father Smyth had cut me short; joined both his hands, and with the stretched fore-finger tapping me on both hands, which were clenched as in prayer, addressed to me these identical remarkable words, "My dear fellow, the licence is a mere *watchword* of the day, and they make a cat's-paw of you."

Mr. Black undertook my defence: *the words above stuck in my throat, though.*

Mr. Hackett, on being consulted, assented that Mr. Rede could promise us to take into consideration the present excited feelings of the diggers, and use his best judgment as to a further search for licences on the morrow.

Mr. Rede: "Yes, yes; but, understand me, gentlemen, I give no pledge."

The usual ceremonies being over, Sub-inspector Taylor kindly escorted us to the bridge, gave the pass-word, and to go — just as any one else will go in this land, who puts his confidence in red-tape — *disappointed.*

XLIII.

LA VOLPE CAMBIA IL PELO, MA NON LA PELLE;
CAMBIA LA PELLE IL SERPE, NON IL VELENO.
IL CANE NON ABBAJA COL VENTRE PIENO;
VESTESI IL LUPO IN PECORA TRA L'AGNELLE.
ANTICA STORIA;
MA SENZA GLORIA.

BY this time, the main road was crowded. The men were anxiously waiting to know our success. Mr. Black calmed their excitement as kindly as circumstances admitted. We returned to our camp at the Eureka. Mr. Black rendered an account of our mission with that candour which characterises him as a gentleman. I wished to correct him in one point only, and said, my impression was, that the Camp, choked with red-coats, would quash Mr. Rede's "good judgment," get the better of his sense, if he had any of either, and that he would come out licence-hunting on an improved style.

Peter Lalor adjourned the meeting to five o'clock in the morning.

XLIV.

ACCINGERE GLADIO TUO SUPER FEMUR TUUM.

ON Friday, December 1st, the sun rose as usual. The diggers came in armed, voluntarily, and from all directions: and soon they were under drill, as the day before. So far as I know, not one digger had turned to work. It may have happened, that certain Cornishmen, well known for their peculiar propensity, of which they make a boast to themselves, to pounce within an inch of their neighbour's shaft, were not allowed to indulge in "encroaching." This, however, I assert as a matter of fact, that the Council of the Eureka Stockade never gave or hinted at any order to stop the usual work on the gold-field.

Towards ten o'clock, news reached our camp that the red coats were under arms, and there would be another licence-hunting.

The flames did not devour the Eureka Hotel with the same impetuosity as we got up our stockade. Peter Lalor gave the order: Vern had the charge, and was all there with his tremendous sword. "*Wo ist der Raffaello! Du, Baricaden bauen,*" and all heaps of slabs, all available timber was soon higgledy-piggledy thrown all round our camp. Lalor then gave directions as to the position each division should take round the holes, and soon all was on the "*qui vive.*"

Had Commissioner Rede dared to rehearse the farce of the riot-act cracking as on Gravel-pits, he would have met with a warm reception from the Eureka boys. It was all the go that morning.

No blue or red coat appeared. — It was past one o'clock: John Bull must have his dinner. Lalor spoke of the want of arms and ammunition, requested that every one should endeavour to procure of both as much as possible, but did certainly *not* counsel or even hint that stores should be *pressed* for it.

A German blacksmith, within the stockade was blazing, hammering and pointing pikes as fast as his thick strong arms allowed him: praising the while his past valour in the wars of Mexico, and swearing that his pikes would fix red-toads and blue pissants especially. He was making money as fast any Yankee is apt on such occasions, and it was a wonder to look at his coarse workmanship, that would hardly stick an opossum, though his pikes were meant for kangaroos and wild dogs.

XLV.

POPULUS EX TERRA CRESCIT: MULTITUDO HOMINUM EST POPULUS; ERGO, MULTITUDO HOMINUM EX TERRA CRESCIT.

BETWEEN four and five o'clock of same afternoon, we became aware of the silly blunder, which proved fatal to our cause. Some three or four hundred diggers arrived from Creswick-creek, a gold-field famous for its pennyweight fortunes — grubbed up through hard work, and squandered in dissipation among the swarm of sly-grog sellers in the district.

We learned from this Creswick legion that two demagogues had been stumping at Creswick, and called the diggers there to arms to help their brothers on Ballaarat, who were worried by scores, by the perfidious hounds of the Camp. They were assured that on Ballaarat there was plenty of arms, ammunitions, forage, and provisions, and that preparations on a grand scale were making to redress once for all the whole string of grievances. They had only to march to Ballaarat, and would find there plenty of work, honour, and glory.

I wonder how honest Mr. Black could sanction with his presence, *such suicidal rant,* such absurd bosh of that pair of demagogues, who hurried down these four hundred diggers from Creswick, helpless, grog-worn, that is, more or less dirty and ragged, and proved the greatest nuisance. One of them, MICHAEL TUEHY, behaved valiantly and so I shall say no more.

Of course something must be done. Thonen was the purveyor. The Eureka butcher on the hill gave plenty of meat, and plenty of bread was got from all the neighbouring stores, and paid for. A large fire was lit in the middle of the stockade, and thus some were made as comfortable as circumstances admitted; others were quartered at the tents of friends; the greater part, soon guessing how they had been humbugged, returned to their old quarters.

Arms and ammunition were our want. Men were there enough; each and all ready to fight: such was the present excitement; but blue and red coats cannot be driven off with fists alone. Lalor gave all his attention to the subject, but would not consent yet to press stores for it.

Vern was perpetually expecting every moment his German Rifle Brigade. Have patience till to-morrow.

In the evening a report was made to the Council, that a reinforcement of soldiers from Melbourne was on the road. Captains Ross and Nealson hastened with their divisions across the bush to intercept the expected troops, so as to get at their arms and ammunition. All proved in vain.

When a revolution explodes as conspired and planned by able leaders, it is usually seen that it was their care from the very beginning, that arms and ammunition should be at hand when and wherever required; while usury, ambition, or vengeance lavishly provide the money to render the revolution popular: but we had never dreamed of making any preparation, because we diggers had taken up arms solely in self-defence; and as up to Saturday the Council of the Eureka Stockade counted in the majority honest men, themselves hard-working diggers, they would not turn burglars or permit anybody to do so in their name.

Truly, I heard from Manning, that a certain committee kept on their hallucinated yabber-yabber at the Star Hotel. I never was there, and know nothing about Star blabs. They, with the exception of Vern, were not with us, thank God; up to Saturday four o'clock any how.

XLVI.

NON IRASCIMINI.

SATURDAY morning. The night had been very cold, we had kept watch for fear of being surprised; every hour the cry, was "The military are coming."

Vern had enlarged the stockade across the Melbourne road, and down the Warrenheip Gully.

Suppose, even that all diggers who had fire arms had been present and plucky, yet no man in his right senses will ever give Vern the credit for military tactics, if that gallant officer had thought that an acre of ground on the surface of a hill accessible with the greatest ease on every side, simply fenced in by a few slabs placed at random, could be defended by a handful of men, for the most part totally destitute of military knowledge, against a disciplined soldiery, backed by swarms of traps and troopers.

Such, however, was our infatuation, that now we considered the stockade stronger, because it looked more higgledy-piggledy.

XLVII.

NON NOBIS, NON NOBIS, SED PAX VOBISCUM.

IT was eight o'clock. Drilling was going on as on the previous day. Father Smyth came inside the stockade: it was my watch. He looked very earnest, a deep anxiety about the hopelessness of our struggle, must have grieved his Irish heart. He obtained permission from Lalor to speak to those under arms, who belonged to his Congregation. Vern consented, and Manning announced it to the men. Father Smyth told them, that the government Camp was under arms, some seven or eight hundred strong; that he had received positive information, that government had sent other reinforcements from Melbourne, which would soon reach Ballaarat; warned them against useless bloodshed; reminded them that they were Christians; and expressed his earnest desire to see all of them at Mass on the following (Sunday) morning.

Father Smyth, your advice was kindly received; if it did not thrive, was it because you sowed it on barren ground?

The following document may in time help to bring forth truth to light: —

"Colonial Secretary's Office,
"Melbourne, lst December, 1854.
"Rev. Sir, —
"In acknowledging the receipt of your letter of yesterday's date, I am desired by his Excellency to thank you for the earnest efforts which, in your professional calling, you are making to allay the disturbances. Unless the government enforce the laws which may be in operation, disorder and licentiousness must prevail.

"You know a commission is issued for the purpose of inquiring into the state and condition of the digging population: until they make their report, the laws his Excellency found in force must be obeyed.

"I have the honour to be, Rev. Sir,
"Your most obedient servant,
"J. MOORE, A.C.S.
"The Rev. Patrick Smyth,
"Catholic Priest, Ballaarat."

XLVIII.

THE THINGS WE ARDENTLY WISH FOR IN THIS LIFE, EITHER NEVER COME TO PASS, OR IF THEY DO IT IS TOO LATE. HENCE, "BETTER LATE THAN NEVER."

THE whole of the morning passed off as quietly as any well wisher to our cause could desire. Towards twelve o'clock it was our decision that licence-hunting was over, for the day any how, since no digger recollected a search for licence taking place on a Saturday afternoon. Our talk was of the coming meeting of the reform league at two o'clock on Sunday, at the Adelphi, as announced at the monster meeting on Wednesday.

The impression was almost general, that "Charley" would soon dismiss the hated brood of our commissioners, and things would then be "all right." "Off to get a bite," was the pass-word.

I assert as a matter of fact, and a living eye-witness, that between one and two o'clock on Saturday, December 2nd, 1854, the Eureka stockade was comparatively deserted. Those who remained (some one hundred) were such, as either had a long distance to go to reach their tents, and the day was very hot, or such as had no tent or friend on Ballaarat. I took notice of this very circumstance from my tent, the second from the stockade, on the hill, west, whilst frying a bit of steak on the fire of my tent chimney, facing said stockade: Manning was peeling an onion. I transcribe the above from the identical note I had taken down on my diary, at the identical hour aforesaid, and can afford to challenge contradiction.

XLIX.

TAEDET ANIMAM MEAM VITÆ MEÆ.

THE news of our private, though never acknowledged, disbandment must soon have reached the Camp.

THE LORD GOD OF ISRAEL UNRAVEL THE MYSTERY.

What a nonsense of mine to endeavour to swell up the Eureka stockade to the level of a Sebastopol!!

Good reader, I have to relate the story of a shocking murder, a disgrace to the Christian name.

I am a Catholic, and believe in the life everlasting. On the day of judgment it will go milder with the Emperor Nicholas, than with the man, whoever he may be, that prompted and counted on the Eureka massacre on the Sunday morning, December 3rd, 1854.

At four o'clock, the diggers crowded again towards the stockade. The divisions of Ross and Nealson had returned from their excursions and were under arms. The scene became soon animated, and the usual drilling was pushed on with more ardour than ever.

John Basson Humffray, of whom nothing was seen or heard since the previous Wednesday, now introduced, through a letter *in his own handwriting*, addressed — "To the Commander-in-Chief of the armed diggers, Eureka," a Doctor Kenworthy, as surgeon, because he (Humffray) feared that a collision between the diggers and the military would soon take place.

Peters, the spy, was at the same time within the stockade.

The "surgeon" had his Yankee face under a bell-top (French hat): he entered into conversation with me in person. I had my sword in hand, and was on watch. We began to talk about MAZZINI and Captain FORBES: this latter, a brave American officer, fought in the late struggle at Rome (1848). I perfectly recollect, that, pointing with a smile to our barricade, I told this Kenworthy, we had thrown them up for our defence against licence-hunting. There is a living witness to the above circumstance, a countryman of mine, whose name I do not remember just now, but he wore at the time a red shirt, with picks and shovels all over it.

Previous to this, Vern, whose silly vanity would by no means allow him to put up with his not having been elected Commander-in-Chief, all on a sudden cried out in his sort of bombast, "Here they are coming, the boys: now I will lead you to death or victory!" — actually a band of men was tramping full speed towards the stockade.

L.

NARRAVERE PATRES NOSTRI
ET NOS NARRAVIMUS OMNES.

WAS it then the long, long-looked for German Rifle Brigade? Here is it's four-horned name — I copy from a slip of paper I wrote in pencil on that very Saturday, as the name was too long and difficult for me to remember — "The Independent Californian Rangers' Revolver Brigade."

I should say they numbered a couple of hundred, looking Californian enough, armed with a Colt's revolver of large size, and many had a Mexican knife at the hip.

Here is the very circumstance when M'Gill made his appearance for the first time within the stockade; I recollect perfectly well the circumstance when a Mr. Smith, of the American Adams's Express, was holding the bridle of the horse, from which said M'Gill dismounted.

JAMES M'GILL is of the breed on the other side of the Pacific. He is thought to have been educated in a military academy, and certainly, he has the manners of a young gentleman of our days. He is rather short, not so much healthy-looking as wide awake. "What's up?" is his motto. This colony will sober him down, and then he will attend more to "what's to be done." His complexion bears the stamp of one born of a good family, but you can read in the white of his eyes, in the colouring of his cheeks, in the paleness of his lips, that his heart is for violence. When he gets a pair of solid whiskers, he may pass for a Scotchman, for he has already a nose as if moulded in Scotland. He speaks the English language correctly, and when not prompted by the audacity of his heart, shows good sense, delicate feelings, a pleasing way of conversation. His honour was impeached by Vern, who never came up to the scratch, though; witness, Mr. John Campbell, of *The Age* office.

When a man is dead, there and then he is himself the horrible evidence of corruption; but, as long as he lives there is hopes for fair play, and hear *his* evidence on the resurrection of life: hence the moral courage to assert the truth, shuts out the physical strength for blather to shampoo the lie; and an honest upright man of education and a Christian leaves *duelering* to fools.

M'Gill is not wicked in heart, though he may not yet have settled-principles. If this world be such a puzzle even for grey-heads, who have seen enough of it, what then must it be for one, come out of College and learning life on the gold-fields? Hence, if I say that he helped with others to draw the chestnuts out of the Eureka Stockade, for some old Fox, I cannot offend him. — Who was the accursed old

Fox? Patience, there is a God. — When I was in gaol, I was not vexed at hearing him at liberty and happy: I could not possibly wish my misery to any one; but his boast on Ballaarat that his friend Dr. Kenworthy had procured him a "written free pardon" did smother me with bitterness.

LI.

TOTA DOMUS DUO SUNT, IIDEM PARENTQUE JUBENTQUE!

A CONFUSION ensued which baffles description; marching, counter-marching, orders given by everybody, attended to by nobody. This blustering concern, when brought forward on the stage at the State Trials, appeared so much to the heart's content of his Honour, of his and my learned friend Mr. Ireland, that I must offer it here, *nolens volens,* for the confirmation of the Cracker-of-high-treason-indictments' approbation.

Thomas Allen examined. — (See Report of the Nigger-Rebel State Trial, in *The Age,* February 24th, 1855.)

"This witness was so very deaf that the Attorney-General had actually to bawl out (*oh! pity the lungs!*) the questions necessary to his examination. He stated, he kept the Waterloo coffee-house and store at the Eureka. He had just returned from Melbourne on the Saturday, December 2nd. He heard inside the stockade the word to 'fall in' for drill. Saw them go through several military evolutions. They did not exactly go through them in a military manner, but in the way in which what call an 'awkward squad' might do. — (I believe you, Old Waterloo; go a-head). He had been at the battle of Waterloo, and knew what military evolutions were. Saw one squad with pikes and another with rifles. He heard one of them say, 'Shoulder poles,' then he said, 'Order poles,' 'Ground arms,' 'Stand at ease,' 'Pick up poles,' 'Shoulder arms,' 'Right face,' 'Quick march,' 'Right counter march,' and they were then marched for more than two hours. After that he saw them 'fall in three deep,' and were then told (by Captain Hanrahan) to prepare to 'receive cavalry,' and 'charge cavalry' — Poke your pike into the guts of the horses, and draw it out from under their tail.

"*After that, in the evening, he saw the man who was in command again form his men around him, and he seemed to be reading a general order for the night.* After it got night, one of them came up to him and said, 'Now, Old Waterloo, you must come and join us,' and he threw down a pike which he told him to take. He said, 'No; it is such a d——d ugly one, I'll have nothing to do with it.' Another came, and witness asked what bounty he gave, saying £50 was little enough for an old Waterloo man. Because he would not join them he was taken into custody by them, and was guarded by three men with pikes at his door. (Great works!) All this was on Saturday. His tent was the second inside the stockade. His tent and all his property was destroyed by fire: it all amounted to £200.

He believed it was set fire to by the police." (And so it was, old Waterloo-no-bolter, good-hearted old man as ever lived in the world. If you wish call for a copy of this book; do.)

NOTICE

GREAT WORKS!!

THIS day, Saturday, November 10th, 1855. A glorious day for Ballaarat: PETER LALOR, our late Commander-in-Chief, being elected by unanimous acclamation, Member of the Legislative Council for this "El Dorado." I spoke at the Camp face to face with James M'Gill. We shook hands with mutual respect and friendship.

M'Gill, at my request, looked full in my eyes, and assured me, that the order old Waterloo speaks of, was to the effect of appointing officers for watch at the stockade, for "out-posts" to keep a sharp look-out, for march to intercept reinforcements; in short, an order for military discipline, very necessary under the prevailing excitement. Said order for the night (Saturday, December 2nd) was drawn up by his command, and written black on white by Alfred, the brother of George Black.

M'Gill further stated that the supposed "Declaration of Independence," on the model of the American one, is a gratuitous falsehood, which must have originated from some well-disposed for, or well-affected to, Toorak small-beer. Hence,

James M'Gill hereby directs me to challenge the production of the document in question, either the original or copy of it, of course with satisfactory evidence of its being a genuine article.

I express the hope that H. R. Nicholls, ex-member of the Local Court, Ballaarat, will take notice of the above.

Let us return to the Eureka stockade.

———————

LII.

QUADRUPEDANTE PUTREM SONITU QUATIT UNGULA CAMPUM.

THE excitement was of Satan. It was reported, the whole of the Melbourne road was swarming with fresh reinforcements. The military would soon attack the stockade, but Vern would lead the diggers to death or glory.

I went out to get positive information, and I did see some two hundred red-coats stationed under arms at the foot of Black hill. The general impression spread like wild-fire that the diggers would now all be slaughtered. I returned, and was anxious to communicate with Lalor. The council room was guarded by Californian faces, perfect strangers to me. The "pass-word" had been changed, and I was refused admittance.

Old colonial-looking fellows rode to and fro from all parts: some brought canisters of gunpowder and bags of shot; others, fire-arms and boxes of caps. They had been pressing stores.

All at once burst out a clamorous shouting. Captain Ross was entering the stockade in triumph with some old fire-arms and a splendid horse. They had been sticking up some three or four tents, called the Eureka government Camp. Great Works! that could have been done long before, without so much fuss about it; and, forsooth, what a benefit to mankind in general, that Commissioner Amos, ever since, was so frightened as to get his large eyes involuntary squinting after his mare!!

Sly-grog sellers got also a little profit out of the Eureka Stockade. A fellow was selling nobblers out of a keg of brandy hanging from his neck. It required Peter Lalor in person to order this devil-send out of the stockade.

"Press for," was the order of the hour. Two men on horseback were crossing the gully below. Young Black — the identical one with a red shirt and blue cap, who took down the names round Lalor's stump, on Bakery-hill on Thursday morning, and who *to the best of my knowledge* never had yet been within the stockade — came out of the committee-room, and hastened up to me with the order to pick out some men and press those two horses in.

I gave him a violent look, and made him understand that "I won't do the bushranger yet." The order was however executed by fresh hands entirely unknown to me, who rushed towards the horsemen, shouted to both of them to stop, and with the threat of the revolver compelled them to ride their horses within the stockade. I felt disgusted at the violence.

The reign of terror will not strike root among Britons, because the Austrian rule does not thrive under the British flag; and so here is a CRAB-HOLE that BRAVE LALOR alone can properly log up.

I asked in German from Vern the "pass-word," and on whispering "Vinegar-hill" to the sentinels, I was allowed to get out of the Stockade.

"Nein, sagte ich mirselbst, nein, eine solche eckliche Wirthschaft habe ich noch nie *geseh'n.*

"Nom d'un nom! c'est affreux. Ces malheureuf sont-ils donc possedes?

"Odi profanum vulgus et arceo.

"Por vida deDios! por supuesto jo fuera el Duke de Alba, esos Gavachos, carajo, yo los pegaria de bueno.

"Che casa del diavolo, per Dio! Che ti pare! niente meno si spalanca l'inferno. Alla larga! Sor Fattorone: Pronti denari, Fan patti chiari. Minca coglione!"

Such were more or less the expressions to give vent to my feelings on my way to the Prince Albert Hotel, Bakery-hill, to meet there a friend or two, especially my old mate, *Adolphus Lessman*, Lieutenant of the Riflemen.

———

LIII.

TURBATUS EST A FURORE OCULUS MEUS.

THE following is the scene, so characteristic of the times, as it was going on at the Prince Albert: —

"Who's the landlord here?" was the growl from a sulky ruffian, some five feet high, with the head of a bull-dog, the eyes of a vulture, sunken in a mass of bones, neglected beard, sun-burnt, grog-worn, as dirty as a brute, — the known cast, as called here in this colony, of a "Vandemonian," made up of low, vulgar manners and hard talk, spiked at each word, with their characteristic B, and infamous B again; whilst a vile oath begins and ends any of their foul conceits. Their glory to stand oceans of grog, joined to their benevolence of "shouting," for all hands, and their boast of black-eye giving, nose-smashing, knocking in of teeth, are the three marks of their aristocracy. Naturally cowards, they have learned the secret that "pluck," does just as well for their foul jobs. Grog is pluck, and the more grog they swallow, the more they count on success. Hence their frame, however robust by nature, wears out through hard drink, and goes the way of all flesh, rarely with grey hairs. It is dangerous to approach them; they know the dodge how to pick up a quarrel for the sake of gratifying their appetite for fighting. You cannot avoid them in this colony; they are too numerous. I saw hundreds of these Vandemonians, during my four months in gaol. Their heart must be of the same stuff as that of vultures, because they are of the same trade. In a word, they are the living witnesses among us, of the terrible saying of Isaiah, "The heart of man is desperately wicked."

Through such did Satan plant his standard to rule this southern land, before Christ could show his CROSS; hence, before famous Ballaarat could point at a barn, and call it a church, on the township, old Satan had three palaces to boast of, the first of which — a match for any in the world — has made the landlord as wealthy and proud as a merchant-prince of the City of London. "*Non ex illis Mecœnates*," — that's the secret how this land has produced so many first-rate bullock-drivers.

The scene at the Prince Albert is now more interesting.

LIV.

IN VINO VERITAS.

THE Vandemonian was, of course, accompanied by nine more of his pals, all of them armed to the teeth with revolvers, swords, pikes, and knives.

Carl Wiesenhavern, a man of noble character, and, therefore a man who hates knavery, and has no fear of a knave, answered with his peculiar German coolness, "Here I am, what do you want?"

"Nobblers round," was the eager reply.

"If that's what you want," replied Wiesenhavern, "you shall have it with pleasure."

"We got no money."

"I did not ask you for any: understand me well, though;" pointing at each of them with the forefinger of his clenched right hand, "you will have a nobbler a-piece, and no more: afterwards you will go your way. Are you satisfied with my conditions?"

"Yes, yes! we agree to that: go on you b——."

Wiesenhavern scorned to notice the fellow, and, according to the old custom of the house, placed two decanters of brandy, together with the tumblers, on the bar, saying, "Help yourselves, gentlemen."

They fell at once upon the brandy, and their mean rascality was shown by some seizing the glass and covering it with the full hand to conceal their greediness.

Nobbler-drinking is an old colonial habit; it gives pluck to the coward when he is "up to something;" so happened it with these fellows.

"Well, landlord, your brandy is d——d good — the real sort of stuff, and no b—y mistake. You shouted nobblers round for all hands — that's all right; it's no more than fair and square now for the boys to shout for you:" and, with a horrible curse, "Fill up the bottles; let's have another round."

Wiesenhavern kept himself quiet. One of the ruffians showed his intention to enter the bar, and play the landlord within. Wiesenhavern coolly persuaded him back by the promise he would fetch from his room, "something rowdy, the right old sort of stuff — *Champagne Cognac, tres vieux.*" The fellows presumed their "bouncing" was all the go now, and laughed and cursed in old colonial style.

Wiesenhavern fetched his pistols, and his partner, Johan Brandt, a double-barrelled gun. Now Mr. Brandt is one of those short, broad-shouldered, sound, dog-headed Germans, with such a determinate look when his otherwise slow wrath is stirred up, that it is not advisable to tackle with his fists, and much less with his rifle. Wiesenhavern, with that precision of manners, which always gains the point on such occasions, placed a decanter full of brandy on the bar, and, with cocked pistols in both hands, said, "Touch it, if you dare; if any one among you got the pluck to put in his tumbler one drop out of that bottle *there*, he is a dead man;" and Mr. Brandt backed him by simply saying: —

"I'll shoot the fellow, like a dog."

What was the result? Of course the same, whenever you deal with knaves — and you make them understand what you mean. They were cowed; and as by this time, the high words had called in several old customers of the house who wished well to it, because they knew it deserved it, so the ruffians had to cut for fear of their own dear lives.

Then it was related with sorrow, that several similar bands were scouring the gold-fields in all directions, and in the name of the committee of the Eureka stockade, under cover of pressing for fire-arms and ammunition, plundered the most respectable stores of all they could lay their hands upon.

One instance, as reported *there* and *then* by parties who had just witnessed the transaction.

A similar gang, four strong, had entered the store of D. O'Conner, on the Golden Point, and asked in the name of the committee, powder and shot, but the vagabonds did not care so much for amunition for their guns, as for the stuff for their guts, what tempted them most was fine good Yorkshire hams, and coffee to wash it down. In short, they ransacked the whole store; and each took care of "something," the best of course, and therefore the cash-box, worth some twenty pounds was not forgotten.

The above are facts. I do not assert that such were the orders of the committee, got up after four o'clock of same Saturday at the Eureka stockade. I had no part or portion in the committee, and know nothing of it personally.

LV.

NON SIT NOBIS VANUM, MANE SURGERE ANTE LUCEM.

I RAN up to the stockade to remonstrate with Peter Lalor, for whom I had too much respect to think for one moment, that he had any hand, and much less that he had sanctioned, such suicidal proceedings.

Thanks to the password; I entered within the stockade. It must have been not far from midnight. I found everything comparatively quiet; the majority were either asleep or warming themselves round the big fire. I spoke in German face to face, for the last time, with Thonen. M'Gill and two-thirds of the Independent Californian Rangers' Rifle Brigade, in accordance with the avocation expressed in the title, were out "starring" to intercept reinforcements reported on the road from Melbourne. Nealson and his division were off for the same purpose. Was their lot that of Lot's wife?

Sir Charles Hotham must have possessed the rod of Moses to convert the quartz of Victoria into red coats, as numerous as the locusts that plagued Pharaoh's land. The Local Court of Ballaarat should recommend His Excellency to carry out the "abolition of shepherding at Sebastopol."

I asked Thonen to see Lalor. I was answered that Peter, from sheer exhaustion, must rest for an hour or two, and was asleep.

Myself not having closed an eye since Thursday, I felt severely the want of sleep. Is not sabbath-keeping our day's cant in the English language? Anyhow it must be admitted, in justice to both silver and gold lace, that they take it in good earnest: to keep the sabbath is a holy and wholesome thing for them. I do not remember what was my frame of mind at the time I wished Thonen good night; very probably, "Enough for the day, the morrow will have its own troubles:" at any rate, Thonen gave word to the "outposts," chiefly Californians to let me pass to my tent: and having thrown myself on my stretcher, with every thing quiet round about, I soon fell asleep.

On the afternoon of Saturday, the following notice was posted up: —

V.   R.

NOTICE.

No light will be allowed to be kept burning in any tent within musket-shot of the line of sentries after 8 o'clock p.m. No discharge of fire-arms in the neighbourhood of the Camp will be permitted for any purpose whatever.

The sentries have orders to fire upon any person offending against these rules.

(By order),   T. BAILEY RICHARDS,
Lieut. 40th Regt., Garrison Adjutant.

LVI.

REMEMBER THIS SABBATH DAY (DECEMBER THIRD), TO KEEP IT HOLY.

I AWOKE. Sunday morning. It was full dawn, not daylight. A discharge of musketry — then a round from the bugle — the command "forward" — and another discharge of musketry was sharply kept on by the red-coats (some 300 strong) advancing on the gully west of the stockade, for a couple of minutes.

The shots whizzed by my tent. I jumped out of the stretcher and rushed to my chimney facing the stockade. The forces within could not muster above 150 diggers.

The shepherds' holes inside the lower part of the stockade had been turned into rifle-pits, and were now occupied by Californians of the I. C. Rangers' Brigade, some twenty or thirty in all, who had kept watch at the "outposts" during the night.

Ross and his division northward, Thonen and his division southward, and both in front of the gully, under cover of the slabs, answered with such a smart fire, that the military who were now fully within range, did unmistakably appear to me to swerve from their ground: anyhow the command "forward" from Sergeant Harris was put a stop to. Here a lad was really courageous with his bugle. He took up boldly his stand to the left of the gully and in front: the red-coats "fell in" in their ranks to the right of this lad. The wounded on the ground behind must have numbered a dozen.

Another scene was going on east of the stockade. Vern floundered across the stockade eastward, and I lost sight of him. Curtain whilst making coolly for the holes, appeared to me to give directions to shoot at Vern; but a rush was instantly made in the same direction (Vern's) and a whole pack cut for Warrenheip.[1]

There was, however, a brave American officer, who had the command of the rifle-pit men; he fought like a tiger; was shot in his thigh at the very onset, and yet, though hopping all the while, stuck to Captain Ross like a man. Should this notice be the means to ascertain his name, it should be written down in the margin at once.

The dragoons from south, the troopers from north, were trotting in full speed towards the stockade.

Peter Lalor, was now on the top of the first logged-up hole within the stockade, and by his decided gestures pointed to the men to retire among the holes. He was shot down in his left shoulder at this identical moment: it was a chance shot, I recollect it well.

A full discharge of musketry from the military, now mowed down all who had their heads above the barricades.

Ross was shot in the groin. Another shot struck Thonen exactly in the mouth, and felled him on the spot.

Those who suffered the most were the score of pikemen, who stood their ground from the time the whole division had been posted at the top, facing the Melbourne road from Ballaarat, in double file under the slabs, to stick the cavalry with their pikes.

The old command, "Charge!" was distinctly heard, and the red-coats rushed with fixed bayonets to storm the stockade. A few cuts, kicks, and pulling down, and the job was done too quickly for their wonted ardour, for they actually thrust their bayonets on the body of the dead and wounded strewed about on the ground. A wild "hurrah!" burst out, and "the Southern Cross" was torn down, I should say, among their laughter, such as if it had been a prize from a May-pole.

Of the armed diggers, some made off the best way they could, others surrendered themselves prisoners, and were collected in groups and marched down the gully. The Indian dragoons, sword in hand, rifle-pistols cocked, took charge of them all, and brought them in chains to the lock-up.

———

1. To chop the gaseous factory of the following electrifying blather, Toorak had offered £500 reward!! Great works.

VERN's LAST LETTER.

(From *The Age*, Monday, January, 15th, 1855.)

The following letter — the last written in these colonies by the now celebrated Vern-has been sent to us for publication. Our readers may rely on its authenticity.

Ship ——, Sydney Heads,
Dec. 24th, 1854.

Farewell to thee, Australia! A few moments more, and then Australia, land of my adoption, adieu! adieu!

Thy rocky shores
Fade o'er the waters blue.

The ship that bears me to exile has spread her wings; but Australia, and you my late companions in arms, I cannot leave you without bidding you (it may be my last) farewell. I part from you, perhaps for ever; but wherever fickle fortune may banish me to, your memory will help to beguile the dreary hours of exile; and I hope that a name once so familiar to you, now an outlaw from injustice and tyranny, may be kindly remembered by you.

. . . .

Oh, that a kind fate had laid me low in the midst of you, and given me a final resting-place, Australia, in thy bosom. But no! Fate denied me a warrior's death, a patriot's grave, and decreed that I should languish in banishment. [*Fate? be d——d: the immoderate length of your legs was fatal to your not getting a "warrior's grave."*]

There was a time when I fought for freedom's cause, under a banner made and wrought by English ladies — [*Ah, ah, I thought you would soon bring in the ladies!* where, please?]

63

Victoria! thy future is bright — [*sweet and smart if Vern be the operator.*] I confidently predict a Bunker's Hill, or an Alma — [*Great works!*] as the issue of your next insurrection. [*No more truck with your legs, though: let's see your signature and be off.*]

Farewell, Australians!

Yours, truly, and for ever,
C.harles H.otham's F.ootman DE LA VERN.

Hold hard, leave us the address where you got your soap last. I want to shampoo my red hair, so as to make my head worth £500. Yankee speculation I guess.

LVII.

DIRIGAT DOMINUS REGINUM NOSTRAM.

THE red-coats were now ordered to "fall in;" their bloody work was over, and were marched off, dragging with them the "Southern Cross."

Their dead, as far as I did see, were four, and a dozen wounded, including Captain Wise, the identical one, I think whom I speak of in relating the events of Tuesday evening, November 28.

Dead and wounded had been fetched up in carts, waiting on the road, and all red-things hastened to Ballaarat.

The following is for the edification of all the well-affected and well-disposed of the present and future generation: —

V.   R.

NOTICE.

Government Camp,
Ballaarat, Dec. 3rd, 1854.

Her Majesty's forces were this morning fired upon by a large body of evil-disposed persons of various nations, who had entrenched themselves in a stockade on the Eureka, and some officers and men killed.

Several of the rioters have paid the penalty of their crime, and a large number are in custody.

All well-disposed persons are earnestly requested to *return to their ordinary occupations, and to abstain from assembling in large groups*, and every protection will be afforded to them by the authorities.

ROBT. REDE,
Resident Commissioner.

God save the Queen.

LVIII.

VERITATEM DICO NON MENTIOR.

HERE begins a foul deed, worthy of devils, and devils they were. The accursed troopers were now within the stockade. They dismounted, and pounced on firebrands from the large fire on the middle of the stockade, and deliberately set in a blaze all the tents round about. I did see with both eyes one of those devils, a tall, thick-shouldered, long-legged, fast Vandemonian-looking trooper, purposely striking a bundle of matches, and setting fire at the corner end, north of the very store of Diamond, where we had kept the council for the defence.

The howling and yelling was horrible. The wounded are now burnt to death; those who had laid down their arms, and taken refuge within the tents, were kicked like brutes, and made prisoners.

At the burning of the Eureka Hotel, I expressed it to be my opinion that a characteristic of the British race is to delight in the calamity of a fire.

The troopers, enjoying the fun within the stockade, now spread it without. The tent next to mine (Quinn's) was soon in a blaze. I collected in haste my most important papers, and rushed out to remonstrate against such a wanton cruelty. Sub-inspector Carter pointing with his pistol ordered me to fall in with a batch of prisoners. There were no two ways: I obeyed. In the middle of the gully, I expostulated with Captain Thomas, he asked me whether I had been made a prisoner within the stockade. "No, sir," was my answer. He noticed my frankness, my anxiety and grief. After a few words more in explanation, he, giving me a gentle stroke with his sword, told me "If you really are an honest digger, I do not want you, sir; you may return to your tent."

Mr. Gordon — of the store of Gordon and M'Callum, on the left of the gully, near the stockade — who had been made prisoner, and was liberated in the same way, and at the same time as myself, was and is a living witness to the above.

On crossing the gully to return to my tent, an infernal trooper trotting on the road to Ballaarat, took a deliberate aim at me, and fired his Minie rifle pistol with such a tolerable precision, that the shot whizzed and actually struck the brim of my cabbage-tree hat, and blew it off my head. Mrs. Davis, who was outside her tent close by, is a living witness to the above.

At this juncture I was called by name from Doctor Carr, and Father Smyth, directed me by signs to come and help the wounded within the stockade.

LIX.

QUIS DABIT CAPITI MEO, AQUAM ET OCULIS MEIS FONTEM LACRYMARUM ET PLORABO DIE AC NOCTE!

I HASTENED, and what a horrible sight! Old acquaintances crippled with shots, the gore protruding from the bayonet wounds, their clothes and flesh burning all the while. Poor THONEN had his mouth literally choked with bullets;[1] my neighbour and mate Teddy More, stretched on the ground, both his thighs shot, asked me for a drop of water. Peter Lalor, who had been concealed under a heap of slabs, was in the agony of death, a stream of blood from under the slabs, heavily forcing its way down hill.

The tears choke my eyes, I cannot write any further.

Americans! your Doctor Kenworthy was not there, as he should have been, according to Humffray's letter.

Catholics! Father Smyth was performing his sacred duty to the dying, in spite of the troopers who threatened his life, and *forced* him at last to desist.

Protestants! spare us in future with your sabbath cant. Not one of your ministers was there, helping the digger in the hour of need.

John Bull! you wilfully bend your neck to any burden for palaver and war to protect you in your universal shopkeeping, and maintain your sacred rights of property; but human life is to you as it was to Napoleon: for him, fodder for the cannon; for you, tools to make money. A dead man needs no further care, and human kind breeds fast enough everywhere after all. — *Cetera quando rursum scribam.*

On my reaching the stockade with a pannikin of water for Teddy, I was amazed at the apathy showed by the diggers, who now crowded from all directions round the dead and wounded. None would stir a finger.

All on a sudden a fresh swarm of troopers cleared the stockade of all moving things with the mere threat of their pistols.

All the diggers scampered away and entered all available tents, crouching within the chimneys or under stretchers. The valorous, who had given such a proof of their ardour in smothering with stones, bats, and broken bottles, the 12th Regiment on their orderly way from Melbourne on Tuesday, November 28, at the same identical spot on the Eureka, now allowed themselves to be chained by dozens, by a handful of hated traps, who, a few days before, had been kept at bay on the whole of the diggings, by the mere shouting of "Joe!" A sad reflection, indeed; a very sad reflection.

Myself and a few neighbours now procured some stretchers, and at the direction of Doctor Carr, converted the London Hotel into an hospital, and took there the wounded.

Said Doctor Carr despatched me to fetch his box of surgical instruments from Dr. Glendinning's hospital on Pennyweight-hill, a distance of a full mile.

I hastened to return, with Dr. Glendinning himself, and I did my best to assist the helpless, and dress their wounds.

___

IMPORTANT. — I must call the attention of my reader to the following fact: — When I entered the stockade with Dr. Carr's surgical box, Mr. Binney, an old acquaintance since the times of Canadian Gully, took me warmly by the hand, and said, "Old fellow, I am glad to see you alive! everybody thinks (pointing to a dead digger among the heap) *that's poor Great Works!*"

The state of mind in which I was, gave me no time to take much notice of the circumstances and must have answered, "Thank God, I am alive," and proceeded to my duty.

The identical Mr. Binney, of the firm of Binney and Gillot, now storekeepers on the Ballaarat township, is a living witness to the above statement.

Solicitor Lynn told me, *in propria persona*, in the Ballaarat prison, that he would take care to bring forward evidence of the above, as he had heard it himself, that such was the case; but I forgot to fee this Lynn, and so he left me to the chance of being "lyn-ched."

1. Carl Wiesenhavern has one of the bullets in his possession.

___

LX.

THE SOUTHERN CROSS, IN DIGGER'S GORE IMBRUED, WAS TORN AWAY, AND LEFT THE DIGGER MOURNING.

THE following Letter, from the able pen of the spirited correspondent of the *Geelong Advertiser* who most undoubtedly must be a digger — that is, one of ourselves, from among ourselves, — is here transcribed as a document confirming the truths of this book: —

THE EUREKA MASSACRE
[From a Correspondent.]
To the Editor of the *Geelong Advertiser* and *Intelligencer.*
Bakery-hill, December 3rd, 1854.

.   .   .   .

Friday you know all about; I will pass that over, and give you a faint outline of what passed under my own eyes. During Saturday, there was a great deal of gloom among the most orderly, who complained much of the parade of soldiery, and the same cause excited a great deal of exasperation in the minds of more enthusiastic persons, who declared that all parties ought to show themselves, and declare whether they were for or against the diggers. Then came a notice from the Camp, that all lights were to be extinguished after eight o'clock, within half-a-mile from the Camp. At this time it was reported that there were two thousand organised men at the Eureka barricade. I was sitting in my tent, and several neighbours dropped in to talk over affairs, and we sat down to tea, when a musket was heard to go off, and the bullet whizzed close by us; I douced the light, and we crept out on our hands and knees, and looked about. Between the Camp and the barricade there was a fire we had not seen before, and occasionally lights appeared to be hoisted, like signals, which attracted the attention of a good many, some of whom said that they saw other lights like return signals. It grew late. TO-MORROW, I FEAR ME, WILL PROVE A DAY OF SORROW, IF THE AFFAIR BE NOT SETTLED BEFORE THEN. I and R—— lay down in our clothes, according to our practice for a week past; and worn out with perpetual alarms, excitement, and fatigue, fell fast asleep. I didn't wake up till six o'clock on Sunday morning. The first thing that I saw was a number of diggers enclosed in a sort of hollow square, many of them were wounded, the blood dripping from them as they walked; some were walking lame, pricked on by the bayonets of the soldiers bringing up the rear. The soldiers were much excited, and the troopers madly so, flourishing their swords, and shouting out — "We have waked up Joe!" and others replied, "And sent Joe to sleep again!" The diggers' Standard was carried by in triumph to the Camp, waved about in the air, then pitched from one to another, thrown down and trampled on. The scene was awful — twos and threes gathered together, and all felt stupefied. I went with R—— to the barricade, the tents all around were in a blaze; I was about to go inside, when a cry was raised that the troopers were coming again. They did come with carts to take away the bodies, I counted fifteen dead, one G——, a fine well-educated man, and a great favourite. [*Here, I think, the Correspondent alluded to me. My friends, nick-named me — Carbonari Great-works.*] I recognised two others, but the spectacle was so ghastly that I feel a loathing at the remembrance. They all lay in a small space with their faces upwards, looking like lead; several of them were still heaving, and at every rise of their breasts, the blood spouted out of their wounds, or just bubbled out and trickled away. One man, a stout-chested fine fellow, apparently about forty years old, lay with a pike beside him: he had three contusions in the head, three strokes across the brow, a bayonet wound in the throat under the ear, and other wounds in the body — I counted fifteen wounds in that single carcase. Some were bringing handkerchiefs, others bed furniture, and matting to cover up the faces of the dead. O God! sir, it was a sight for a sabbath morn that, I humbly implore Heaven, may never be seen again. Poor women crying for absent husbands, and children frightened into quietness. I, sir, write disinterestedly, and I hope my feelings arose from a true principle; but when I looked at that scene, my soul revolted at such means being so cruelly used by a government to sustain the law. A little terrier sat on the breast of the man I spoke of, and kept up a continuous howl: it was removed, but always returned to the same spot; and when his master's body was huddled, with the other corpses, into the cart, the little dog jumped in after him, and lying again on his dead master's breast, began howling again.

—— was dead there also, and ——, who escaped, had said, that when he offered his sword, he was shot in the side by a trooper, as he was lying on the ground wounded. He expired almost immediately. Another was lying dead just inside the barricade, where he seemed to have crawled. Some of the bodies might have been removed — I counted fifteen. A poor woman and her children were standing outside a tent; she said that the troopers had surrounded the tent, and pierced it with their swords. She, her husband, and children, were ordered out by the troopers, and were inspected in their night-clothes outside, whilst the troopers searched the tent. Mr. Haslam was roused from sleep by a volley of bullets fired through his tent; he rushed out, and was shot down by a trooper, and handcuffed. He lay there for two hours bleeding from a wound in his breast, until his friends sent for a black-smith, who forced off the handcuffs with a hammer and cold chisel. When I last heard of Mr. Haslam, a surgeon was attending him, and probing for the ball. R——, from Canada, [*Captain Ross, of Toronto, once my mate*] escaped the carnage; but is dead since, from the wounds. R—— has effected his escape. [*Johnny Robertson, who had a striking resemblance to me, not so much in size as in complexion and colour of the beard especially. Poor Johnny was shot down dead on the stockade; and was the identical body which Mr. Binney mistook for me. Hence the belief by many, that I was dead.*] V—— is reported to be amongst the wounded [*Oh! no, his legs were too long even for a Minie rifle*]. One man was seen yesterday trailing along the road: he said he could not last much longer, and that his brother was shot along-side of him. All whom I spoke to were of one opinion, that it was a cowardly massacre. There were only about one hundred and seventy diggers, and they were opposed to nearly six hundred military. I hope all is over; but I fear not: for amongst many, the feeling is not of intimidation, but a cry for vengeance, and an opportunity to meet the soldiers with equal numbers. There is an awful list of casualties yet to come in; and when uncertainty is made certain, and relatives and friends know the worst, there will be gaps that cannot be filled up. I have little knowledge of the gold-fields; but I fear that the massacre at Eureka is only a skirmish. I bid farewell to the gold-fields, and if what I have seen is a specimen of the government of Victoria, the sooner I am out of it the better for myself and family. Sir, I am horrified at what I witnessed, and I did not see the worst of it. I could not breathe the blood-tainted air of the diggings, and I have left them for ever.

You may rely upon this simple statement, and submit it if you approve of it, to your readers.

I am, Sir.

———

———

LXI.

AB INITIO USQUE AD FINEM HORRIBLIE DICTU.

AVANIT IL TUO COSPETTO, DIO POTENTE! GRIDA VENDETTA IL SANGUE INNOCENTE.

I. Document.

As I want to be believed, so I transcribe the following document from *The Argus* of Friday, December 15th, 1854. — Gordon Evans, one of H. M. Captains in the Eureka massacre, now acts in the capacity of magistrate! —

DEPOSITION OF HENRY POWELL.

"The deceased deposed to the following effect: — My name is Henry Powell. I am a digger residing at Creswick-creek. I left Creswick-creek about noon on Saturday, December 2nd. I said to my mates, 'You'll get the slabs ready. I will just go over to see Cox and his family at Ballaarat.' I arrived at Ballaarat about half-past four, or thereabouts. I saw armed men walking about in parties of twenty or thirty; went to Cox's tent; put on another pair of trowsers, and walked down the diggings. Looked in the ring (the stockade). After that, went home, went to bed in the tent at the back of Cox's tent, about half-past nine. On Sunday morning, about four, or half-past, was awoke by the noise of firing. Got up soon after, and walked about twenty yards, when some trooper rode up to me. The foremost one was a young man whom I knew as the Clerk of the Peace. He was of a light, fair complexion, with reddish hair. He told me to 'stand in the Queen's name! You are my prisoner.' I said 'Very good, Sir.' Up came more troopers. I cannot say how many. Believe about twenty or thirty. I said, 'Very well, gentlemen (!) don't be in a hurry, there are plenty of you,' and then the young man struck me on the head with a crooked knife, about three feet and a half long, in a sheath. I fell to the ground. They then fired at me, and rode over me several times. I never had any hand in the disturbance. There, that's all.

Ballaarat, Dec. 11, 1854.

FIRST CASE of an inquest which has taken place since the massacre of the memorable 3rd. The evidence as to the murder of Powell (*writes* The Argus *express correspondent*) is but a specimen of the recitals heard on every hand of the reckless brutality of the troopers that morning.

VERDICT OF THE JURY.

The death of deceased, Henry Powell, gold-digger, was caused by sabre cuts and gun shot wounds, *wilfully* and *feloniously*, and of their malice *aforethought* inflicted and fired by ARTHUR PURCELL AKEHURST, Clerk of the Peace, Ballaarat bench, and other persons unknown.

The jury return a verdict of *Wilful Murder* against *A. P. Akehurst* and other persons unknown.

The jury express their condemnation of the conduct of Captain Evans, in not swearing deceased at the time of taking his statement after having been cautioned by Dr. Wills of his immediate danger. The jury view with extreme horror the brutal conduct of the mounted police in firing at and cutting down unarmed and innocent persons of both sexes, at a distance from the scene of disturbance, on December 3rd, 1854.

WILLIAMS, Coroner.

Mind, good reader, the above is a legal document.

After my trial, on my way to Ballaarat, I met in Geelong the identical Akehurst, cracking some nuts with (I mean, speaking to) *some young ladies.*

I DESPAIR OF THIS COLONY.

May it please HER MAJESTY to cause inquiry to be made into the character of such that have branded the miners of Ballaarat as disloyal to their QUEEN.

LXII.

TEMPORA NOSTRA.

THE following documents are put in here as evidence of "our times."
II. Document.

BALLAARAT.
THE STATEMENT OF FRANK ARTHUR HASLEHAM.
(Now lying wounded at Ballaarat.)

"Whereas I, Frank Arthur Hasleham, a native of the good town of Bedford, and son of a military officer, to wit, William Gale Hasleham, who bore His Majesty's commission in the 48th Foot at Talavera, and afterwards retired from the 6th veteran battallion:

"Whereas I, the aforesaid, having, in my capacity of newspaper correspondent at Ballaarat, shown, on all proper occasions in general, so especially during the late insurrectionary movement here, a strong instinctive leaning to the side of law, authority, and loyalty, was, on the morning of the 3rd instant, fired at and wounded at a time when the affray was over, and the forces with their prisoners were on the point of returning to the Camp, and in a place whence the scene of action was invisible, and when no other bloodshed had taken place;

71

"On these considerations I desire to make on oath the following statements of facts, as they occurred, and as witnessed by others: —

"Shortly after daybreak in the morning mentioned, my three mates and myself were aroused from sleep by the fire of musketry, a great proportion of the balls whistling over our tents. The tent is pitched on a rising ground about 500 yards south of the stockade; the tent and stockade, each situated on an eminence, are separated by a large gully running east and west, and comprising in its breadth nearly the whole of the distance above specified. Considerably alarmed at the continuance of the firing, we at last got up and went outside, thinking to find a place of shelter of comparative security. After I had gone outside the firing gradually fell off, the stockade was unoccupied, the insurgents' flag was struck, and whatever fighting was then going on was confined to the further slope of the hill on which the stockade was situated. As some desultory firing was still going on, I advanced about fifty yards down the gully, in order to insure safety by getting upon lower ground; by this time, with the exception of an occasional cheer from the military or police, everything was perfectly quiet, and from where I stood neither soldier nor trooper was to be seen. A few minutes after a small detachment of mounted police made its appearance on the hill, and drew up in a line on the either side of the stockade, the officer in command appeared to be haranguing them. I was standing about three hundred yards from them, several other people being near at hand. I saw three troopers leave the ranks and advance towards me; when one of them who rode considerably a-head of the other two arrived within hailing distance, he hailed me as a friend. Having no reason to think otherwise of him, I walked forward to meet him. After he had lured me within safe distance, namely about four paces, he levelled his holster pistol at my breast and shot me. Previous to this, and while advancing towards each other, he asked me if I wished to join his force; I told him I was unarmed, and in a weak state of health, which must have been plain to him at the time, but added that I hoped *this madness on the part of the diggers would soon be over*, upon that he fired."

The trooper be d——d; but I congratulate poor Frank, of the good town of Bedford, for "this madness on the part of the diggers" procuring him £400 sterling from Toorak; so that he can afford to spare me the trouble of encroaching any further into his "statement." Great works!

III. Document more important, by far.

On the 28th November, when some military and ammunition came on the ground, the detachment was set on at Eureka, near the site of the stockade, and in the hubbub consequent the troops were somewhat at fault, and the officer in command called at the London Hotel to inquire the way to the Camp. The owner of the hotel, Mr. Hassall, on being asked, came out of his establishment to point out the way to the officer in command of the detachment, while so doing he received a ball in his leg, and was for a while laid up by the wound. After a long time of suffering, and a great loss of money directly and indirectly, he applied for

compensation — with what success may be seen from the following letter just
come to hand: —

———

"Colonial Secretary's Office,
"Melbourne, 26th October, 1855.
"Gentlemen. — The memorial of the miners on behalf of Mr. B. S. Hassall,
wounded during the disturbances at Ballaarat, having been by the governor's
directions referred to the board appointed to investigate such claims, the
board reported, that from the evidence, it appears impossible Mr. Hassall
could have received his wound from the military, and that they could not see
anything to justify their recommending any compensation for him. His
Excellency cannot therefore entertain the petition as he has not power to
award compensation except on the recommendation of the board.
"I have the honour to be, gentlemen,
"Your most obedient servant,
"J. MOORE, A.C.S."
"SAMUEL IRWIN, and
A. C. BRUNNING, Esqrs."

"Great works" this time at Toorak, eh! oh! dear.
So far so good, for the present; because spy "Goodenough" wants me in the next
chapter.

———

LXIII.

ET SCIAS QUIA NIHIL IMPIUM FECERIM.

IT was now between eight and nine o'clock. A patrol of troopers and traps stopped
before the London Hotel. Spy Goodenough, entered panting, a cocked pistol in his
hand, looking as wild as a raven. He instantly pounced on me as his prey, and
poking the pistol at my face, said in his rage, "I want you."
"What for?"
"None of your d——d nonsense, or I shoot you down like a rat."
"My good fellow dont you see? I am assisting Dr. Carr to dress the wounds of my
friends!" — I was actually helping to bandage the thigh of an American digger,
whose name, if I recollected it, I should now write down with pleasure, because he
was a brave fellow. He had on his body at least half-a-dozen shots, all in front, an
evident proof, he had stood his ground like a man.
Spy Goodenough would not listen to me. Dr. Carr spoke not a word in my
behalf, though I naturally enough had appealed to him, who knew me these two
years, to do so. This circumstance, and his being the very first to enter the stockade,
after the military job was over, though he had *never before* been on the Eureka during
the agitation, his appointment to attend the wounded diggers that were brought up

to the Camp, and especially his absence at my trial, were and are still a *mystery* to me. I was instantly dragged out, and hobbled to a dozen more of prisoners outside, and we were marched to the Camp. The main road was clear, and the diggers crawled among the holes at the simple bidding of any of the troopers who rode at our side.

LXIV.

SIC SINUERUNT FATA.

ON reaching the Camp, I recognized there the identical American Kenworthy. I gave him a fearful look. I suspected my doom to be sealed.

The soldiers were drinking *ad libitum* from a pannikin which they dipped into a pail-bucket full of brandy. I shall not prostitute my hand, and write down the vile exultations of a mob of drunkards. It was of the ordinary colonial sort, whenever in a fight the "ring" is over.

Inspector Foster, commanded us to strip to the bare shirt. They did not know how to spell my name. I pulled out a little bag containing some Eureka gold-dust, and my licence; Mr. Foster took care of my bag, and just as my name was copied from my licence; a fresh batch of prisoners had arrived and Mr. Foster was called outside the room where I was stripping. Now, some accursed trooper pretended to recognize me as one of the "spouts" at the monster meeting. I wanted to keep my waistcoat on account of some money, and papers I had in the breast pocket; my clothes were literally torn into rags. I attempted to remonstrate, but I was kicked for my pains, knocked down in the bargain, and thrown naked and senseless into the lock-up.

The prison was crammed to suffocation. We had not space enough to lie down, and so it was taken in turns to stand or lie down. Some kind friend sent me some clothes, and my good angel had directed him to bury my hand-writings he had found in my tent, under a tent in Gravel-pits.

Fleas, lice, horse-stealers, and low thieves soon introduced themselves to my notice. This vermin, and the heat of the season, and the stench of the place, and the horror at my situation, had rendered life intolerable to me. Towards midnight of that Sunday I was delirious. Our growls and howling reached the Commissioner Rede, and about two o'clock in the morning the doors were opened, and all the prisoners from the Eureka stockade, were removed between two files of soldiers to the Camp store-house a spacious room, well ventilated and clean. Commissioner Rede came in person to visit us. Far from any air of exultation, he appeared to me to feel for our situation. As he passed before me, I addressed him in French, to call his attention to my misery. He answered very kindly, and concluded thus: —

"*Je ne manquerai pas de parler au Docteur Carr, et si ce que vous venez de me dire se trouve vrai,* je veux bien m'interesser pour vous."

"*Vous etez bien bon, Monsieur le Commissionaire,*" repondis-je.

Il faut donc que j'aie eu des ennemis bien cruels au Camp! Avaient-ils soif de mon sang, ou etaient-ils de mercenaires? Voila bien un secret, et je donnerai de coeur ma vie pour le percer. Dieu leur pardonne, moi, je le voudrais bien! mais je ne saurai les pardonner jamais.

LXV.

ECCE HOMO.

ON Monday morning, the fresh air had restored me a little strength. We had an important arrival among us. It was the Editor of *The Times* newspaper, arrested for sedition. All silver and gold lace, blue and red coats in the Camp rushed in to gaze on this wild elephant, whose trunk it was supposed, had stirred up the hell on Ballaarat.

HENRY SEEKAMP is a short, thick, rare sort of man, of quick and precise movements, sardonic countenance; and one look from his sharp round set of eyes, tells you at once that you must not trifle with him. Of a temper that must have cost him some pains to keep under control, he hates humbug and all sort of yabber-yabber. His round head of tolerable size, is of German mould, for the earnestness of his forehead is corrected by the fullness of his cheeks, and a set of moustachios is the padlock of his mouth, whose key is kept safe in his head, and his heart is the turn-key. When his breast is full, and he must make it clean, its gall will burn wherever it falls, and set the place a blazing. To keep friends with such a cast of mind, whose motto is Nelson's, you must do your duty; never mind if you sink a shicer, bottom your shaft any how. You are his enemy if you are or play the flunky; he will call you a "thing," and has a decided contempt for "incapables." Hence, his energy was never abated, though the whole legion of Victorian red-tape wanted to dry his inkstand, and smother his lamp in gaol. That there are too many fools at large, he knows, because he has travelled half the world; what he can not put up with, is their royal cant, religious bosh, Toorak small-beer, and first and foremost, their money-grubbing expertness. Hence, now and then, his ink turns sour, and thereby its vitriol burns stronger. *The Times,* of which he is the founder, is the Overseer of Ballaarat, and the *Dolce far niente* will not prosper.

Our literary prisoner was literally insulted, and could not look with enough contempt on all those accursed asses braying (at him) "*The Times!*" "*The Times!*"

I felt for him very much, and joined conversation with him in French. I state it as a matter of fact, that *there* and *then* I had the presentiment that all the spies pointed me out there, and only there and then as his accomplice. Towards ten o'clock we were ordered to fall in, in four rows. Now the Camp officials and their myrmidons were in their glory. They came to number their prey, and mark out a score of heads to make an "example" of, for the better conduct of future generations. Unfortunately for my red hair, fizzing red beard, and fizzing red moustachios, my name was taken down after the *armed ruffian* and the *anonymous scribbler,* and followed by that of the *nigger-rebel.*

It was odious to see honourable, honest, hard-working men made the gazing stock of a parcel of pampered perverted fools, for the fun of "a change;" to gratify their contempt for the blue-shirt and thick boots who had dared, mucky and muddy, to come out of their deep wet holes to hamper these gods of the land in their dog's game of licence-hunting!

LXVI.

THEN the following document was shown for our edification: —

VICTORIA GOVERNMENT GAZETTE EXTRAORDINARY.
(Published by Authority.)

MARTIAL LAW

DECLARED IN THE DISTRICT OF BUNINYONG.

PROCLAMATION

By his Excellency Sir Charles Hotham, Knight-Commander of the Most Honourable Military Order of the Bath, Lieutenant-Governor of Victoria, &c., &c., &c.

WHEREAS bodies of armed men have arrayed themselves against Her Majesty's forces and the constituted authorities, and have committed acts of open rebellion: and whereas, for the effectual suppression thereof it is imperatively necessary that Martial Law should be administered and executed within the limits hereinafter described: Now I, the Lieutenant-Governor of the said Colony with the advice of the Executive Council thereof: do hereby command and Proclaim that MARTIAL LAW from and after twelve of the clock at noon on Wednesday, the sixth day of December instant, shall and may be administered against every person and persons within the said limits, who shall at any time after the said hour commit any act of rebellion, any treason, treasonable or seditious practices, or other outrage or misdemeanor whatsoever within the following limits, that is to say: Yarrowee . . . Lal Lal . . . Moorabool . . . Ran Rip . . . Yarrowee aforesaid. And I do hereby, with the advice aforesaid, order and authorize all officers commanding Her Majesty's forces to employ them with the utmost vigour and decision for the immediate suppression of the said rebellion and offences, and to proceed against and punish every person and persons acting, aiding, or in any manner assisting in the said rebellion and offences, according to Martial Law, as to them shall seem expedient for the punishment of all such persons: And I do hereby especially declare and proclaim, that no sentence of death shall be carried into execution against any such person without my express consent thereto: [*Great works!*] And I do hereby with the advice aforesaid, notify this my Proclamation to all subjects of Her Majesty in the Colony of Victoria.

Given under my Hand and the Seal of the Colony, at Melbourne, this fourth day of December, in the year of Our Lord one thousand eight hundred and fifty-four, and in the eighteenth year of Her Majesty's Reign.

(L.S.)   CHARLES HOTHAM,

By His Excellency's Command,

JOHN FOSTER.

God save the Queen!

Great works!

LXVII.

ECCE AMARITUDO MEA AMARISSIMA.

WE were frightened by the report that a gang of red-coats were sinking a large pit in the Camp.

"Are they going to bury us alive without any flogging? That's not half so merciful as Haynau's rule in Austria;" was my observation to a mate prisoner — a shrewd Irishman.

"Where did you read in history that the British Lion was ever merciful to a fallen foe?" was his sorrowfully earnest reply.

Oh! days and nights of the 3rd, 4th, 5th, and 6th of December, 1854, your remembrance will not end, no, not even in my grave.

They were happy days in my youth, when I thought with Rousseau, that the heart of man is from nature good. It was a sad fatality now that compelled me to feel the truth from the prophet Isaiah, that the heart of man is desperately wicked.

I was really thunderstruck at the savage eagerness with which spies and red-coats sprang out of their ranks to point me out. Though a British soldier was not ashamed to swear and confess his cowardice of running away from before my pike, which I never had on the stockade, where the fellow never could have seen me; I shall not prostitute my intelligence and comment on the "evidence" against me from a gang of bloodthirsty mercenary spies. The printer will copy my trial from the public newspaper, *The Age*.

LXVIII.

CONDEMN THE WICKED, AND BRING HIS WAY UPON HIS HEAD, OH, LORD GOD OF ISRAEL!

THE first witness against me was such a rum sort of old colonial bird of the jackass tribe, and made such a fool of himself for Her Majesty's dear sake, about the monster meeting, where as it appeared, he had volunteered as reporter of the

Camp; that now God has given him his reward. He is a gouty cripple, still on "Her Majesty's fodder" at the Camp, Ballaarat.

Who will sharpen my quill and poison my inkstand, that I may put to confusion the horrible brood of red-tape that ruled on Ballaarat at the time. To administer justice in the sacred name of Her Most Gracious Majesty, they squandered the sweat of self-over-working diggers, on a set of devils, such that they actually competed with one another, in vomiting like sick dogs! Their multitude was taken as a test of their veracity, on the Mosaical ground, that "out of the mouth of two witnesses shall the guilty be condemned;" and yet, with the exception of spy Goodenough, and spy Peters, *none other* to my knowledge ever did see my face before.

I assert and declare as an honest man and a Christian, that my eyes never did see the witnesses against me, before I was under arrest at the Camp. My soul was drowned in an ocean of bitterness when of that brood of Satan, one did swear he had run from before my pike; another had fired at me, but his pistol "snapped;" a third made me prisoner within the stockade; a fourth took me up chained to other prisoners who had surrendered, from the stockade to the Camp.

Such, then, is the perversity of the human heart! In vain did I point out to the sitting magistrate the absurdness of their evidence, and the fact that Sub-inspector Carter, and Dr. Carr could prove the contradiction. I was so embittered and broken-hearted at the wickedness of so many infuriated mercenary rascals, that had made up their mind to sell the blood of an honest man, in as much as I had repeatedly told each and all of them, when they came to "recognize" in our prison, that they must mistake me for another as I was not within the stockade that Sunday morning; that I . . . but it is too humiliating to say any more.

Mr. Sturt, with an odious face, whose plumpness told me at once he was no friend to fasting, strutted to the magisterial chair, and committeed me and the nigger-rebel, to whom I was kindly hobbled, to take our trial for high treason!

LXIX.

VOX POPULI, VOX DEI.

IN the course of the day (December 7th), in spite of all the bayonets and blunderbusses, the report reached us that the Melbourne people had had a Monster Meeting of their own, equal to ours of November 29th, and that Mr. Foster, the "Jesuit," had been dismissed from office.

The tragical act on Ballaarat was over; the scenery was changed; and the comedy now proceeded to end in the farce of the State Trials in Melbourne.

Between Wednesday and Thursday, all the 160 prisoners were liberated, with the necessary exception of thirteen, reserved to confirm the title of this book.

I do not wish to omit one significant circumstance. On Tuesday night, December 5th, I was hobbled for the night to young Fergusson, an American, and shared with him his blankets. I felt very much for this young man, for he suffered from consumption; and as I do respect him, so I shall not disclose our private conversation. This, however, is to the purpose. He was among us, and with us at four o'clock on Saturday, at one and the same time when spy Peters was within the stockade.

No spy, no trap, no trooper appeared against young Fergusson. Doctor Kenworthy, his countryman, had the management of getting him off. I was glad at his obtaining his liberty, because he was a brave, kind-hearted, republican-minded young American, and I intend to keep his blue blankets he left to me in prison for my comfort, in his remembrance.

LXX.

AUDI ALTERAM PARTEM.

"FAIR PLAY."

AS I wish to be believed, so I transcribe the following from *The Argus*, Friday, December 15, 1854;

MAINTENANCE OF LAW AND ORDER.

The Lieutenant-Governor received a deputation from, with an address signed by, five hundred bankers, merchants, and other classes resident in Melbourne, placing themselves, their services, and their influence unreservedly at His Excellency's disposal, for the maintenance of order and upholding of the paramount authority of existing (!) law.

His Excellency listened with marked attention to the address, to which he gave the following answer: —

"Gentlemen . . .

. . . "It is necessary to look its (the Colony's) difficulty full in the face.

"Here we have persons from all parts of the globe — men who come to look for gold and gold alone; men of adventurous spirit, of resolution, and of firm purpose to carry out the principles which actuate them. If gold fails, or the season is unfavourable, we must expect such outbreaks and such dangers as have given rise to the most loyal and valuable address which you present to me. [*Pardon, Monseigneur, apres lecture des versets 28, 29, du chap. I., et versets 17, 18, 19, du chap. III., de la Genese, favorisez s'il vous plait l'exploitation de l'activite de tous ces gaillards la, par la Charrue: il n'y a pas mal de terres ici, et bien pour tout le monde. Audaces fortuna juvat.*]

"I desire to govern by the people, and through the people: and by the people I mean through the intelligence of the people. [*Elle est fameuse, Monseigneur l'intelligence de ceux, qui vous ont conseille l'affaire de Ballaarat! surtout la farce odieuse de haute-trahison!*]

"In Ballaarat it was not a particular law, against which objection was raised, nor was there a particular complaint made. [*Oh, pardon, Monseigneur: ou l'on vous a toujours mal informe; ou l'on vous a souvent cache la verite: malheureusement, cela n'a pas beaucoup change meme aujourd'hui.* Vide *The Times*, Ballaarat, Saturday, September 29, 1855, and Saturday, November 10th — Local Court.]

... "It was not exactly the licence fee, that caused the outbreak, though that was made the '*nom de guerre,*' the '*cheval de bataille,*' this was not the real cause. I consider that the masses were urged on by designing men who had ulterior views, and who hoped to profit by anarchy and confusion. [*Comment se fait-il Monseigneur que vous mettez le prix de £500 sur la tete du chef de ces blagueurs du Star Hotel, a Ballaarat; et puis vous lassiez courir le malin a son aise! Avez-vous, oui ou non, Monseigneur, accorde votre pardon a M'Gill? et les autres Americains donc?*]

"Then we have active, designing, intriguing foreigners, who also desire to bring about disorder and confusion." [*Cependant, moi, bon garcon apres tout, et d'une ancienne famille Romaine, j'ai ete VOLE sous arret au Camp de Ballaarat par VOS gens et avec impunite, Monseigneur. Vous me faites l'honneur d'avouer par votre lettre la chose, mais vous n'avez point fait de restitution. Ce n'est pas comme cela que j'entends le vieux mot Anglais, Fair-play.*]

Hence, I had better address myself to the five hundred gentlemen, who belong to the brave Melbourne people after all.

Gentlemen,

Five hundred copies of this work, which costs me an immense labour, for the sake of the cause of truth, will be left with

MESSRS. MUIR, BROTHERS AND CO.,

Merchants, Flinders-lane, Melbourne — of the same firm much respected on Ballaarat, to whom I am personally known long ago, having been their neighbour on the Massacre-hill, Eureka. Ten shillings is my price for each copy: and, as Messrs. Muir render this service to me gratuitously, so I hereby authorise them to keep half-a-crown from each ten shillings, and in the spirit of St. Matthew, verses 1, 2, 3, 4, chap. vi., share said half-crowns in the following proportion: one shilling to the Benevolent Asylum; one shilling to the Melbourne Hospital, and sixpence to the Miners' Hospital, Ballaarat.

I hope thus to understand "Fair-play" better than Toorak.

I have not yet done with His Excellency's answer.

"The part which the bankers, merchants, tradesmen and others in Melbourne and in Geelong [*pas a Ballaarat, Monseigneur*], have taken in coming forward to support me, I shall be careful to represent properly at home, where perhaps these occurrences may attract more attention than they deserve. [*Pour votre bonheur, Monseigneur, Sebastopol leur donne assez d'occupation pour le moment.*]

"I shall declare my opinion that the mass of the community does not sympathise with these violators of the law." [*Est-ce donc un reve, Monseigneur, que votre gouvernment en voulait a ma tete, aussi, bien qu'a celle de douze autres prisonniers d'etat, et que le peuple nous a acquitte glorieusement par*

SEVEN BRITISH JURIES!]

Mon ardent desir, mon tourment presque, c'est d'avoir vite l'honneur de parler, encore une fois sur la terre, a

SA MAJESTE LA REINE VICTORIA.
AINSI-SOIT-IL.

LXXI.

THE STATE PRISONERS.

I BEG to say at once, that with the exception of Hayes and Manning, of the remaining ten, seven were perfect strangers to me; three I had simply met at work on the gold-fields; and I won't say anything further.

Yes, though, MICHAEL TUHEY was the stoutest heart among us, an Irishman in word and deed, young, healthy, good-hearted chap, that hates all the ways of John Bull, he had been misled by honest George Black countenancing the two demagogues at Creswick-creek, and had hastened with his double-barrelled guns to Ballaarat, and stood his ground like an Irishman, against the red-coats. He never was sorry for it. His brother paid some forty pounds to a certain solicitor for his defence, but when Mic was tried for his neck, the Hog was not there. GOD SAVE THE PEOPLE!

THOMAS DIGNAM, a serious-looking, short, tight-built young chap, a native of Sydney, who hated all sort of rogues, because he was honest in heart. He brunted courageously the mob fury on Tuesday evening, November 28th, on the Eureka, and actually saved at the risk of his own life, the life of a soldier of the 12th regiment on its way to Ballaarat; he took up arms in the cause of the diggers in Thursday's licence-hunt, was subsequently under drill at the stockade; fought like a tiger on Sunday morning; repented not of having put on stretchers a couple of red-coats; was always cheerful, contented and kind-hearted during the four months in gaol; paid his last farthing out of the honest sweat of his brow, to Stephens his solicitor for the defence (above thirty pounds) and when put in the dock to take his trial for high-treason, lo! there was no charge against him; the prosecution was dropped. GOD SAVE THE QUEEN!

We are however still in chokey at Ballaarat. We were put under the officious care of Sergeant Harris, who condescended to show some affection for Joseph, to prove that his Christian love could extend even to niggers; but the red-coat wanted to draw worms from the black rebel. We were nigh bursting for laughter, when Joseph during his two days' trial came into our yard for his meals, and related to us with such eye-twinklings, widening of nostrils, trumping up the lips, scratching all the while his black wool so desperately, and the doodle music of his unearthly whistle! *"how old chappyman and a tother smart 'un of spin-all did fix that there mob of traps; 'specially that godammed hirpocrit of sergeant, I guess."*

JOHN JOSEPH, a native of New York, under a dark skin possessed a warm, good, honest, kind, cheerful heart; a sober, plain-matter-of-fact contented mind; and that is more than what can be said of some half-a-dozen grumbling, shirking, snarling, dog-natured state prisoners.

Sergeant Harris took it into his head to humble Hayes — humility is also a Christian virtue — and so honoured him with the perfumery job of clearing the tub at the corner, full of urine and solids. Hayes, for the lark did it once, but found it against his principles to practise on said tub again, and thus got into disgrace with our over-kind sergeant.

To be serious: Timothy Hayes, our chairman at the monster meeting, aristocratically dressed among us, had like the rest his plump body literally bloated with lice from the lock-up. Poor Manning was the worst. Myself, I was plagued with that disgusting vermin, all through those ignominious four months in gaol.

It were odious to say many, many other things.

LXXII.

IS THERE A MORTAL EYE THAT NEVER WEPT?

ON Sunday afternoon, we witnessed a solemn scene, which must be recorded with a tear wherever this book may find a reader.

The sun was far towards the west. All had felt severely the heat of the day. The red-coats themselves, that were of the watch, felt their ardour flagging. Of twelve prisoners, some gazed as in "a fix," and were stationary; others, "a-cursing," swept up and down the prison; the rest, cast down, desponding, doing violence to themselves, to dam their flooded eyes. I was among the broken-hearted.

MRS. HAYES, who in the days of her youth must have made many young Irish hearts ache "for something," had brought now a bundle of clean clothing, and a stock of provisions, to make her husband's journey to Melbourne as comfortable as possible. There she was, holding her baby sucking at her breast; her eyes full on her husband, which spoke that she passionately loved him. Six children, neatly dressed, and the image of their father, were around. Timothy Hayes forced himself to appear as cheerful as his honourable heart and proud mind would allow. He pressed his little daughter, who wanted to climb his shoulder; he pronounced his blessing on the younger of his sons. The eldest (twelve years old) was kissing his father's left hand, bathing it all the while with such big, big tears, that dropped down so one by one, and so one after the other!

Good boy, your sorrows have begun soon enough for your sensible heart! Strengthen it by time with Christian courage, or else you will smother it with grief, long before your hair has turned grey! There are too many troubles to go through in this world. Take courage; there is a God, and therefore learn by heart the Psalm, "*Beatus vir qui timet Dominum.*" My head has still the red hair of my youth, and yet I am a living witness of many truths in that Psalm; meditate, therefore, especially on the last verse, ending "*Desiderium peccatorum peribit.*"

Had I in younger years cultivated painting, I feel satisfied that I could produce now such a tableau as to match any of my countryman, *Raffaelle*; so much an all-wise Providence has been pleased, perhaps for the trial of my heart, to endow me with a cast of mind that, on similar occasions as the solemn one above, whenever my electric fluid is called into action, it is actually a daguerreotype.

LXXIII.

AMARE RIMEMBRANZE.

AT four o'clock on Tuesday morning, we were commanded to fall in, dressed and hobbled as we were. Captain Thomas, with the tone and voice of a country parson, read to us his "Order of the day," to the effect that we were now under his charge for our transit to Melbourne; that if any of us stirred a finger, or moved a lip — especially across the diggings — his orders were that the transgressor should be shot on the spot. This arrangement, so Austrian-like, and therefore unworthy of a British officer, did not frighten us, and I cried, loud enough, "God save the Queen!"

Inspector FOSTER sprang up to me with his hopping leg, put on me tighter darbies, and together with the mulatto-rebel put us in front of the cart, giving strict orders to shoot us both down if we attempted to turn our heads. *Veritatem dico, non mentior;* and so Messrs. Haynau, Jellachich, and Co., from that morning my hatred for you is on the decline.

They rode us through the main road as fast is it was safe for the preservation of our necks — the only thing *they* wanted to preserve inviolate for head-quarters.

Though it was clear daylight, yet I did see only one digger on the whole of the main road.

On passing through the Eureka, I got a glance of my snug little tent, where I had passed so many happy hours, and was sacred to me on a Sunday. There it lay deserted, uncared for! My eyes were choked with tears, and at forty years of age a man does not cry for little.

LXXIV.

DELLA VITA LO SPELLO DAL MONDO SCIOLTO,
AL MONDO VIVO PERCHE NON SON SEPOLTO.

WE were soon in Ballan. Good reader, please enter now within my mind. The lesson, if read, learned, and inwardly digested, will be of good use for the future. The troubles of this colony have begun.

It is eight o'clock of a fine morning; the spring season is in its full: the sun in his splendour is all there on the blue sky. Nature all around is life. The landscape is superb. It reminded me *della Bella Cara Itallia.* The bush around was crammed with parrots, crows, and other chattering birds of the south. They were not prevented from singing praises each in its own language to the Creator, and all was joy and happiness with them. Unfortunately those lands lay uncultivated by the hand of man; but were not left idle by nature. Lively, pretty little flowers of the finest blue, teemed here, there, and everywhere, through the splendid grass, wafted to and fro by a gentle wind.

Look now at the foot of the picture.

There were *thirteen* of us all healthy, honest, able-bodied men, chained together on three carts. A dozen of dragoons, strong, sound-looking men, were riding on horseback as sharp-shooters, in all directions, before our carts in the bush. Their horses were really splendid animals. A score of troopers of the accursed stamp we had then on Ballaarat, sword unsheathed, carbines cocked, kept so close to our carts that one of these Vandemonians was half jammed on riding by a large gum-tree; was thrown from his horse, and disabled, but not killed. We are at last in Ballan, for change of horses. Captain Thomas and a stout healthy-looking man, with a pair of the finest black whiskers I ever saw, in the garb of a digger, who gave such orders to the coachman, as were always attended to, with the usual colonial oaths as a matter of course, were regaling themselves with bottled porter on a stump of a tree outside the public-house. The dragoons and troopers had biscuit, cheese, and ale served to them, though paid for by themselves, before our teeth.

There was no breakfast for the poor state prisoners, in chains, and lying on the bare ground. They had some trouble before they could obtain from the red-coats watching over them, and blowing heaps of smoke from stump pipes, *a drop of cold water* — I mean actually a drop of cold water.

Good reader, you know WHOM I did bless, *whom* I did curse.

LXXV.

PETITE, SED NON ACCIPIETIS, QUIA PETISTIS.

THE following document, which does honour and justice to its writer, J. Basson Humffray, to the 4500 of our fellow-miners of Ballaarat, who signed it, to the state prisoners themselves, is now here transcribed as necessary to the purpose of this book.

THE BALLAARAT DELEGATES, AND THEIR INTERVIEW WITH HIS EXCELLENCY SIR CHARLES HOTHAM, K.C.B., &c.

The public has already seen the written reply of His Excellency to the petition from Ballaarat, signed by nearly 4500 of the inhabitants of that important, but "officially" ridden place.

We deem it our duty to the public, and especially to those whose delegates we are, to state the main reasons urged by us for a general amnesty, and to make some general remarks thereon, and also upon the reply. We have delayed doing this, as we expected to have returned immediately to Ballaarat, and we did not wish to forestall our intended statement at a public meeting, which would have been held on our return; but as circumstances interfere with this arrangement, we now give our report.

We were very kindly and respectfully received by His Excellency. We thought it right to state that we repudiated physical force as a means of obtaining constitutional redress, believing that the British constitution had sufficient natural elasticity to adapt itself to the wants of the age, and would yield under proper pressure. But the arming of the diggers of Ballaarat, however reprehensible it might have been in itself, claims to be judged on special grounds, inasmuch as they had special provocation. The diggers of Ballaarat were attacked by a military body under the command of civil (!) officers, for the production of licence-papers, and, if they refused to be arrested, deliberately shot at. The diggers did not take up arms, properly speaking, against the government, but to defend themselves against the bayonets, bullets, and swords of the insolent officials in their unconstitutional attack, who were a class that would disgrace any government, by their mal-administration of the law.

The diggers did not take up arms against British rule, but against the *mis*-rule of those who were paid to administer the law properly; and however foolish their conduct might be, it was an ungenerous libel on the part of one of the military officers to designate *outraged British subjects* as "foreign anarchists and armed ruffians."

The diggers were goaded on to take the stand they did by the "digger-hunt," of the 30th November, which, we are sustained in saying, was a base piece of gold and silver lace revenge. Facts will no doubt appear by-and-bye, elucidating and confirming this statement.

We reminded His Excellency of the fact, that the public had asked for or sanctioned a general amnesty; and although we were prepared to admit that it was unbecoming the dignity of any government to give way to what was termed "popular clamour," yet, in this case, the good and the wise amongst all classes, forming a very large proportion of the inhabitants, had asked for it, and we thought the general wish should not be *lightly treated*. His Excellency observed, "Certainly not." We argued that an amnesty would restore general confidence, and secure support to the government in any emergency; and, even supposing there was any one in the movement who sought to overturn the government, instead of overturning corruption, and establishing a better system of administration, a general amnesty would silence such, as the great majority of the diggers were content to live under British law, if properly administered; and every one knows there has been much to condemn in the administration of the laws, on the Ballaarat gold-fields especially; and we endeavoured to impress upon the mind of the Lieutenant-Governor, that it was equally true that the majority of those who were proud of being British subjects, were growing tired of waiting for simple justice. And if the executive wish to secure their confidence and support, they must give better evidence of their good intentions of making better laws, or laws better suited to the wants of the people, and securing "equal justice to all." Their recent conduct has created disaffection amongst the ranks of the best disposed; in fact, those who disapproved of the resort to arms on the part of the diggers, condemn in the most unqualified manner the conduct of the Ballaarat officials in collecting a tax (obnoxious at the best) at the bayonet's point, and of the late Colonial Secretary, who could unblushingly write to Commissioner Rede (who superintended the digger-hunt on the 30th of November, and, no doubt, counselled the Sunday morning's butchery), thanking him for his conduct on those occasions! And that if His Excellency would allow us to strip the matter of its official colouring, he would see things in a very different light than they had been officially represented.

That an amnesty would not only secure the confidence of the people in the Governor, but it would show the confidence of the Governor in the people — it would be looked upon as a proof of the strength and vigour of the British constitution, instead of weakness in those that administer the laws under its guidance.

That His Excellency could well afford to be generous.

That, in asking for an amnesty, we were aware it was asking for much, and what a statesman should not do without due deliberation. But at the same time, we submitted we did not ask anything inconsistent with the true interests of the colony, or derogatory to the dignity and honour of the throne itself.

That a general amnesty to the state prisoners would tend much to consolidate the power of the British government in this colony, and show that the representative of Majesty here can afford to be just — to be generous; with the full confidence that such an act would meet with the full concurrence of the Queen of England, and the approbation of the whole British empire. That in this he would act wiser far in listening to the voice of the people than to the short-sighted counsel of the law-advisers of the Crown. Humanity has higher claims than the mere demands and formalities of human law.

We forbear saying all that might be said as to the spies being sent from the Camp to enrol themselves amongst the insurgents, and who, report says, *urged them to attack the Camp*, which was repudiated by the diggers — they saying they would act upon the defensive.

That we believed the enforcement of the law in this case would have the most pernicious effect, not only upon the commerce of the colony, but would retard, if not prevent, the accomplishment of those schemes of reform that His Excellency had promised.

That if he valued the good opinions of the people — the peace and prosperity of the colony, he would be giving the best evidence of it by granting the amnesty we prayed for; but that, if His Excellency punished these men, it would be calling into existence an agitation which would, we feared, end in civil commotion, if not in the disseverance of the colony from the mother country.

That we thought there were reasons sufficiently important to justify an amnesty, on the grounds of state policy alone.

But even supposing there were no legitimate grounds for an amnesty, and that the government have been right in all that they have done — which would be saying what facts do not warrant — surely the slaughter of some fifty people is blood enough to expiate far greater crimes than the diggers of

Ballaarat have been guilty of, without seeking the lives of thirteen more victims. The government would act wisely in not pursuing so suicidal a course.

His Excellency states, in his written reply, that the diggers, notwithstanding his promise of inquiry into all their grievances, had forestalled all inquiry.

On this head, we would wish to remark, that the fault lies at the door of the government, in prostituting the military, by making them tax collectors, and placing them at the disposal of a few vain officials, who were not over-stocked with brains, and ignorant of the functions of constitutional government. But one fact they seemed fully sensible of, viz.: That "Othello" occupation would indeed soon "be gone," and they were determined to "crush the scoundrels" who dared to question the policy, or even justice, or a government keeping up such an expensive army of La Trobian idlers as strut about in borrowed plumes with all the insolence of office; who, in fact, have proved themselves, with a few honourable exceptions, fit for little else than bringing the colony into debt; creating disaffection amongst the people, and stamping indelible disgrace upon any government that would uphold the system that tolerates them. One of these "retiring" gentlemen stated on the morning of the famed "digger-hunt" of the 30th November, in reply to one of the refractory diggers: "If you do not pay your licences, how are we to be supported at the Camp?" and further, "There are some disaffected scoundrels I am determine to arrest!" To crush! for what? For daring to refuse to pay taxes except they had a voice in the expending of them for the public weal; public taxes are public property. Some of these "gentlemanly" officials made use of language on the occasion alluded to, that not only gave evidence of considerable malignity, but of a vulgarity that a gentleman would scorn to use; and we think it not an unfair inference to draw from the foregoing facts, that the digger-hunt of the 30th of November, and the cruel slaughter of the 3rd December, were unmistakable acts of petty official revenge; and, therefore, instead of the diggers forestalling the Commission of Inquiry, appointed by His Excellency, we advisedly say it was Commissioner Rede and Co. who forestalled the inquiry by endeavouring to crush the "500 scoundrels" he complained of — a scoundrel in that gentleman's estimation seems to be one who thinks that some £12 per head per annum is rather too heavy a tax for an Englishman to pay; especially if used in supporting men so unfit for office as he has proved himself to be. This gentleman was the arch-rioter of the 30th of November; in this we are confirmed (if confirmation of well-known facts were needed) by the verdict of acquittal of the so called " Ballaarat Rioters," partially on the evidence of Mr. Rede himself.

In the latter part of His Excellency's reply, he very properly lays it down as "the duty of government to administer equal justice to all;" which is no doubt the noblest principle of the English constitution, and we certainly have no fears for the peace of even colonial society, with all its supposed discordant elements, so long as that principle is practically carried out; but we are under well founded apprehension if the reverse is to be the order of the day.

There is a paragraph in our petition to the effect, that if "His Excellency had found sufficient extenuation in the conduct of American citizens," we thought there were equally good grounds for extending similar clemency to all, irrespective of nationality; and that it was unbecoming the dignity of any government to make such exceptions; and if such have been done (and that something tantamount to it has been done, there is ample proof), it is a violation of the very principle enunciated by His Excellency in his report viz., "That it is the duty of a government to administer equal justice to all." What we contend for is this: — If it be just to grant an amnesty to a citizen of one country, "equal justice" claims an amnesty for all. We wish it to be distinctly understood by our American friends, that we do not for a moment find fault with His Excellency for allowing their countrymen to go free, but we do complain, in sorrow, that he does not display the same liberality to others — that he does not wisely and magnanimously comply with the prayer of our petition by granting a general amnesty.

But is is stated further in the reply, that "no exception had been made in favour of any person against whom a charge was preferred." With all becoming deference to His Excellency, we think this does not meet the point. If the gentleman were innocent, why guarantee him against arrest? And if his friends (and we give them credit for good tact) anticipated the "preferment of a charge," it does not create any special grounds for an amnesty in contradistinction to a general amnesty.

Again, upon whom lies the *onus* of "preferring a charge?" £500 was offered for *Vern*, "DEAD OR ALIVE" and £400 for Lalor and Black; and yet we presume there was no charge, or charges, "preferred" against them any more than the gentleman alluded to. We yet trust that the same good feeling that induced His Excellency to give James M'Gill his liberty will increase sufficiently strong to unbar the prison-doors, and set the state captives free, that they may be restored to their homes, their sorrowing families, and sympathising countrymen. By such an act, the Lieutenant-Governor will secure the peace of society, and the respect and support of the people, and be carrying out the glorious principle he has proclaimed of "Equal Justice to All."

J. BASSON HUMFFRAY,
C. F. NICHOLLS,
(of Ballaarat.)

Melbourne, 23rd January, 1855.

LXXVI.

QUID SUM MISER, NUNC DICTURUS.

AT Bacchus Marsh we were thrown into a dark lock-up, by far cleaner than the lousy one of Ballaarat. Captain Thomas, who must have acknowledged that we had behaved as men, sent us a gallon of porter, and plenty of damper; he had no occasion to shoot down any of us. I write now this his kindness with thanks.

At last, after a long, long day, smothered with dust, burning with thirst, such that the man in the garb of a digger had compassion on us, and shouted a welcome glass of ale to all of us — we arrived before the Melbourne gaol at eight o'clock at night.

From the tender mercies of our troopers, we were given up to the gentle grasp of the turnkeys. The man in the garb of a digger introduced us to the governor, giving such a good account of us all, that said governor, on hearing we had had nothing to eat since mid-day, was moved to let us have some bread and cheese.

We were commanded to strip to the bare shirt — the usual ignominy to begin a prison life with — and then we were shown our cell — a board to lie down on, a blanket — and the heavy door was bolted on us.

Within the darkness of our cell, we now gave vent to our grief, each in his own way.

Sleep is not a friend to prisoners, and so my mind naturally wandered back to the old spot on the Eureka.

LXXVII.

REQUIESCANT IN PACE.

LALOR's Report of the Killed and Wounded at the Eureka Massacre, on the morning of the memorable Third of December, 1854: —

The following lists are as complete as I can make them. The numbers are well known, but there is a want of names. I trust that the friends or acquaintances of these parties may forward particulars to *The Times* office, Ballaarat, to be made available in a more lengthened narrative.

KILLED.

1 JOHN HYNES, County Clare, Ireland.
2 PATRICK GITTINS, Kilkenny, do.
3 —— MULLINS, Kilkenny, Limerick, Ireland.
4 SAMUEL GREEN, England.
5 JOHN ROBERTSON, Scotland.
6 EDWARD THONEN (lemonade man), Elbertfeldt, Prussia.
7 JOHN HAFELE, Wurtemberg.
8 JOHN DIAMOND, County Clare, Ireland.
9 THOMAS O'NEIL, Kilkenny, do.
10 GEORGE DONAGHEY, Muff, County Donegal, do.
11 EDWARD QUIN, County Cavan, do.
12 WILLIAM QUINLAN, Goulbourn, N.S.W.
13 and 14 Names unknown. One was usually known on Eureka as "Happy Jack."

WOUNDED AND SINCE DEAD.

1. LIEUTENANT ROSS, Canada.
2 THADDEUS MOORE, County Clare, Ireland.
3 JAMES BROWN, Newry, do.
4 ROBERT JULIEN, Nova Scotia.
5.——CROWE, unknown.
6 ——FENTON, do.
7 EDWARD M'GLYN, Ireland.
8 No particulars.

WOUNDED AND SINCE RECOVERED.

1 PETER LALOR, Queen's County, Ireland.
2 Name unknown, England.
3 PATRICK HANAFIN, County Kerry, Ireland.
4 MICHAEL HANLY, County Tipperary, do.
5 MICHAL O'NEIL, County Clare, do.
6 THOMAS CALLANAN, do. do.
7 PATRICK CALLANAN, do. do.
8 FRANK SYMMONS, England.
9 JAMES WARNER, County Cork, Ireland.
10 LUKE SHEEHAN, County Galway, do.
11 MICHAEL MORRISON, County Galway, do.
12 DENNIS DYNAN, County Clare, do.

(Signed)   PETER LALOR,
Commander-in-Chief.

What has become of GEORGE BLACK, was, and is still, a MYSTERY to me. I lost sight of him since his leaving for Creswick-creek, on December 1, 1854.

LXXVIII.

HOMO NATUS DE MULIERE, BREVI VIVENS TEMPORE REPLETUR MULTIS MISERIIS. QUI QUASI FLOS CONTERRITUR ET EGREDITUR; POSTEA VELUT UMBRA DISPERDITUR.

IT is not the purpose of this book, to begin a lamentation about my four long, long months in the gaol. My health was ruined for ever: if that be a consolation to any one; let him enjoy it. To say more is disgusting to me and would prove so to any one, whose motto is "Fair-play."

A dish of hominy (Indian meal), now and then fattened with grubs, was my breakfast.

A dish of scalding water, with half a dozen grains of rice, called soup, a morsel of dry bullock's flesh, now and then high-flavoured, a bit of bread eternally sour — any how the cause of my suffering so much of dysentery, and a couple of black murphies were my dinner.

For tea, a similar dish of hominy as in the morning, with the privilege of having now and then a bushranger or a horse-stealer for my mess-mate, and often I enjoyed the company of the famous robbers of the Victoria Bank.

But the Sunday! Oh the Sunday! was the most trying day. The turnkeys, of course, must enjoy the benefit of the sabbath cant, let the prisoners pray or curse in their cells. I was let out along with the catholics, to hear mass. I really felt the want of Christian consolation. Our priest was always in a hurry, twice did not come, once said half the mass without any assistant; never could I hear two words together out of his short sermon. Not once ever came to see us prisoners.

After mass, I returned to my cell, and was let out again for half an hour among all sorts of criminals, some convicted, some waiting their trial, in the large yard, to eat our dinner, and again shut up in the cell till the following Monday.

LXXIX.

"SOUVENIRS" DE MELBOURNE.

FIVE things I wish to register: the first for shame; the second for encouragement; the third for duty; the fourth for information; the fifth for record.

1. We were one afternoon taken by surprise by the whole gang of turnkeys, ordered to strip, and subjected to an ignominious search. The very private parts were discovered and touched. *Veritatem dico, non mentior.*

2. Manning felt very much the want of a chew of tobacco. He and Tuhey would make me strike up some favourite piece out of the Italian opera, and the charm succeeded. A gentle tap at the door of our cell was the signal to get from a crack below a stick of tobacco, and then we were all jolly. We decreed and proclaimed that even in hell there must be some good devils.

3. Mr. Wintle, the governor, inclining to the John Bull in corporation, had preserved even in a Melbourne gaol, crammed as it is at the end of each month with the worst class of confirmed criminals, his good, kind heart. With us state prisoners, without relaxing discipline, he used no cruelty — spoke always kindly to us — was sorry at our position, and wished us well. He had regard for me, on account of my bad health; that I shall always remember.

4. Some day in January we received a New-Year's Present — that is a copy of the indictment. I protest at once against recording it here: it is the coarsest fustian ever spun by Toorak Spiders. I solemnly declare that to my knowledge the name of Her Most Gracious Majesty was never mentioned in any way, shape, or form whatever, during the whole of the late transactions on Ballarat. I devoured the whole of the

indictment with both my eyes, expecting to meet with some count charging us with riot. The disappointment was welcome, and I considered myself safe. Not so, however, by a parcel of shabby solicitors. They said it would go hard with any one if found guilty. The government *meant* to make an example of some of us, as a lesson to the ill-affected, in the shape of some fifteen years in the hulks. They had learned from Lynn of Ballaarat that there were no funds collected from the diggers for the defence. *Cetera quando rursum scribam* — and thus they won some £200 out of the frightened state prisoners, who possessed ready cash.

"What will be the end of us, Joe?" was my question to the nigger-rebel.

"Why, if the jury let us go, I guess we'll jump our holes again on the diggings. If the jury won't let us go, then" — *and bowing his head over the left shoulder, poking his thumb between the windpipe and the collarbone, opened wide his eyes, and gave such an unearthly whistle, that I understood perfectly well what he meant.*

LXXX.

THE STATE PRISONERS.

(From *The Age*, February 14th, 1855.)

THE following is the copy of a letter addressed by the state prisoners now awaiting their trial in the Melbourne Gaol, to the Sheriff, complaining of the treatment they have received: —

"Her Majesty's Gaol, Melbourne,
"February 6th, 1855.

"To the Sheriff of the Colony of Victoria: —

"SIR — As the chief officer of the government, regulating prison discipline in Victoria, we, the undersigned Ballaarat state prisoners, respectfully beg to acquaint you with the mode of our treatment since our imprisonment in this gaol, in the hope that you will be pleased to make some alterations for the better.

"At seven o'clock in the morning we are led into a small yard of about thirty yards long and eight wide, where we must either stand, walk or seat ourselves upon the cold earth (no seats or benches being afforded us), and which at meal times serves as chair, table, etc., with the additional consequence of having our food saturated with sand, dust, and with every kind of disgusting filth which the wind may happen to stir up within the yard.

"We are locked in, about three o'clock in the afternoon, four or five of us together, in a cell whose dimensions are three feet by twelve, being thus debarred from the free air of heaven for sixteen hours out of the twenty-four. The food is of the very worst description ever used by civilized beings. We are debarred the use of writing materials, except for purposes of pressing necessity; are never permitted to see a newspaper; and strictly prohibited the use of tobacco and snuff. We have been subjected to the annoyance of being stripped naked, a dozen men together, when a process of 'searching' takes place that is debasing to any human being, but perfectly revolting to men whose sensibilities have never been blunted by familiarity with crime — an ordeal of examination, and the coarse audacity with which it is perpetrated, as would make manhood blush, and which it would assuredly resent, as an outrage upon common decency, in any other place than a prison. And again, when the visiting justice makes his rounds, we are made to stand bareheaded before him, as if — etc.

"We give the government the credit of believing that it is not its wish we should be treated with such apparent malignity and apparent malice; and also believe that if you, sir, the representative of government in this department, had been previously made acquainted with this mode of treatment, you would have caused it to be altered. But we have hitherto refrained from troubling the government on the subject, in expectation of a speedy trial, which now appears to be postponed *sine die*.

"We, each of us, can look back with laudable pride upon our lives, and not a page in the record of the past can unfold a single transgression which would degrade us before man, or for which we would be condemned before our Maker. And we naturally ask why we should be treated as if our lives had been one succession of crime, or as if society breathed freely once more at being rid of our dangerous and demoralising presence. Even the Sunday, that to all men in Christendom is a day of relaxation and comparative enjoyment, to us is one of gloom and weariness, being locked up in a dreary cell from three o'clock Saturday evening till seven on Monday morning (except for about an hour and a half on Sunday); thus locked up in a narrow dungeon for forty consecutive hours! We appeal to you, and ask, was there ever worse treatment, in the worst days of the Roman inquisition, for men whose reputation had never been sullied with crime?

"We therefore humbly submit, that, as the state looks only at present to our being well secured, we ought to be treated with every liberality consistent with our safe custody; and that any unnecessary harshness, or arrogant display of power, is nothing more or less than wanton cruelty.

"Some of us, for instance, could wile away several hours each day in writing, an occupation which, while it would fill up the dreary vacuum of a prison life, as would the moderate use of snuff and tobacco cheer it, and soothe that mental irritation consequent upon seclusion. But that system of discipline which would paralyse the mind and debilitate the body — that would destroy intellectual as well as physical energy and vigour, cannot certainly be of human origin.

"Trusting you will remove these sources of annoyance and complaint,
"We beg to subscribe ourselves,
"Sir,
"Your obedient servants."
[Here follow the names.]

Sheriff CLAUDE FARIE, Inspector PRICE, Turnkey HACKETT, they will praise your names in hell!

LXXXI.

QUEM PATRONEM ROGATURUS.

THE brave people of Melbourne remembered the state prisoners, forgotten by the Ballaarat diggers, who now that the storm was over, considered themselves luckily cunning to have got off safe; and therefore could afford to "joe" again; the red-streak near Golden-point, having put every one in the good old spirits of the good old times.

> YOURSELF devoting to the public cause,
> You ask the people if they be "there" to die:
> Yes, yes hurrah the thund'ring applause,
> Too soon, alas! you find out the lie!
> Cast in a gaol, at best you are thought a fool,
> Red hot grows your foe; your friend too cool.

An angel, however, was sent to the undefended state prisoners. Hayes and myself were the first, who since our being in trouble, did grasp the hand of a gentleman, volunteering to be our friend.

JAMES MACPHERSON GRANT, solicitor, is a Scotchman of middle-size, middle-height; and the whole makes the man; an active man of business, a shrewd lawyer, and up to all the dodges of his profession. His forehead announces that all is sound within; his benevolent countenance assures that his heart is for man or woman in trouble. He hates oppression; so say his eyes. He scorns humbug; so says his nose. His manners declare that he was born a gentleman.

I very soon gave him hints for my defence, quite in accordance with what I have been stating above, and his clerk took the whole down in short-hand. He encouraged me to be of good cheer, "You need not fear," said he, "you will soon be out, all of you."

God bless you, Mr. Grant! For the sake of you and Mr. Aspinall, the barrister, I smother now my bitterness, and pass over all that I suffered on account of so many postponements.

Timothy Hayes, when we returned broken-hearted for the FIFTH (!) time to our gaol, did we not curse the lawyers!

A wild turn of mind now launched my soul to the old beloved spot on the Eureka, and there I struck out the following anthem.

LXXXII.

VICTORIA'S "SOUTHERN CROSS."

Tune — The "Standard Bearer."

I.

WHEN Ballaarat unfurled the "Southern Cross,"
Of joy a shout ascended to the heavens;
The bearer was Toronto's Captain Ross;
And frightened into fits red-taped ravens.

Chorus. For brave Lalor —
Was found "all there,"
With dauntless dare:
His men inspiring:
To wolf or bear,
Defiance bidding,
He made them swear —
Be faithful to the Standard, for victory or death. (*Bis.*)

II.

Blood-hounds were soon let loose, with grog imbued,
And murder stained that Sunday! Sunday morning;
The Southern Cross in digger's gore imbrued,
Was torn away, and left the diggers mourning!

Chorus.

III.

Victoria men, to scare, stifle, or tame,
Ye quarter-deck monsters are too impotent;
The Southern Cross will float again the same,
UNITED Britons, ye are OMNIPOTENT.

Chorus.

Thus I had spanned the strings of my harp, but the strain broke them asunder in the gaol.

LXXXIII.

INITIUM SAPIENTIE EST TIMOR DOMINI.

THERE are circumstances in life, so inexplicable for the understanding; so intricate for the counsel; so overwhelming for the judgment; so tempting for the soul; so clashing with common sense; so bewildering for the mind; so crushing for the heart; that even the honest man cannot help at moments to believe in FATE. Hence the "*sic sinuerunt Fata,*" will dash the fatalist a-head, and embolden him to knock down friend or foe, so as to carry out his conceit. If successful, he is a Cæsar; if unsuccessful, ignominy and a violent grave are the reward of his worry.

If this be true, as far is it goes, whilst

> Through living hosts and changing scenes we rove,
> The mart, the court, the sea, the battle-plain,
> As passions sway, or accident may move;

it holds not true in a gaol. There you must meet yourself, and you find that you are not your God. Hence these new strings in my harp.

TO THE POINT.

I.

> GAY is the early bloom of life's first dawn,
> But darker colours tinge maturer years;
> Our days as they advance grow more forlorn,
> Hope's brightest dreams dissolve away in tears
> Which were the best, *to be or not to have been?*
> The question may be asked, no answer can be seen.

II.

> On earth we live, within our thoughts — the slaves,
> Of our conceptions in each varied mood,
> Gay or melancholy; — it is the waves

Of our imaginings, become the food
The spirit preys upon; and laughs or raves
With madness or with pleasure, as it would
If drunk with liquids. WE EXIST AND DWELL
AS THE MIND MAY DISPOSE, IN HEAVEN OR IN HELL.

THEME.

Death which we dread so much, is but a name.

SONNET.

He who never did eat his bread in tears;
    Who never passed a dreary bitter night,
    And in his bed of sorrow, the hard fight
Of pending troubles saw, with anxious fears:
Who never an exile forlorn for years,
    And never wept with Israel "at the sight
    Of the waters of Babylon" (Psalm 137), the might
Of Heaven's word is unknown to his ears.
IS THERE A MORTAL EYE THAT NEVER WEPT?
    WITH *tears* the child begins his wants to show
In *tears* the man out of the earth is swept.
    Whether we bless or grumble here below,
HIM who ever in His hand the world has kept
In dark affliction's school we learn to know.

(Of course my original is in Italian.)

————

LXXXIV.

JUDICA ME DEUS, ET DISCARNE CAUSEM MEAM DE GENTE NON SANCTA; AB HOMINE INIQUO ET DOLOSO ERUE ME.

SUPREME COURT
Melbourne, Victoria, Australia Felix,
Wednesday, March 21st, 1855.
(Before his Honour Mr. Justice Barry.)

MY STATE TRIAL

HIS HONOUR took his seat shortly after ten o'clock. The prisoner, that is myself, was placed in the dock, and the following Jury sworn (after the usual challenging):—

PHILLIP BRAGG, Gore-street, Farmer,
ALEXANDER BARTHOLOMEW, Brighton-road, Joiner,
JAMES BLACK, Greville-street, Butcher,
CHARLES BUTT, Lennox-street, Carpenter,
THOMAS BELL, Lennox-street, Carpenter,
FREDERICK BAINES, Richmond-road, Painter,
CHARLES BELFORD, Kew, Gardener,
WILLIAM BROADHURST, Wellington-street, Grocer,
JOSEPH BERRY, Hawthorne, Farmer,
DAVID BOYLE, Kew, Gardener,
WILLIAM BARNETT, Heidelberg, Gardener,
JOHN BATES, Rowena-street, Baker.

Brava gente. Dio vi benedica. Mio Fratello desidera veder ciascuno di Voi, nella nostra Bella Itallia.

For the first time in my life (37 years old), I was placed in a felon's dock, and before a British jury.

The first glance I gave to the foreman made me all serene. I was sure that the right man was in the right place.

JAMES MACPHERSON GRANT, my attorney for the defence, was "all there."

RICHARD DAVIS IRELAND, barrister, my counsel, was heavy with thunder. Thick, sound, robust, round-headed as he is, the glance of his eyes is irresistible. A pair of bushy whiskers frame in such a shrewd forehead, astute nose, thundering mouth; that one had better keep at a respectful distance from drakes. His whole head and strong-built frame tell that he is ready to settle at once with anybody; either with the tongue or with the fist. His eloquence savours pretty strongly of Daniel O'Connell, and is flavoured with colonial pepper; hence Mr. Ireland will always exercise a potent spell over a jury. If he were the Attorney-General, the colony would breath freer from knaves, rogues, and vagabonds. The "sweeps," especially, could not possibly prosper with Ireland's pepper.

According to promise, another lawyer, a man of flesh, had to be present: but, as he was not there, so he is not *here*.

Mr. ASPINALL, barrister, totally unknown to me before, volunteered his services as my counsel to assist Mr. Ireland.

"*In memoria eterna manet amicus*" BUTLER COLE ASPINALL. The print of generous frankness in your forehead, of benevolence in your eyes, of having no-two-ways in your nose, of sincere boldness in your mouth; your height, fine complexion, noble deportment, indicate in you the gentleman and the scholar. If now and then you fumble among papers, whilst addressing the jury, that is perhaps for fear it should be observed that you have no beard; in order that proper attention may be paid to your learning, which is that of a grey-headed man; and though it may be said, that the Eureka Stockade was hoggledy enough, yet your *pop, pop, pop,* was also doggledy.

You know a tree by its fruits; and so you may know, if you like, the Attorney-General by his High-Treason Indictment. I have not the patience to go through it a second time. There are too many Fosters, fostering and festering in this Victorian land.

JUDGE BARRY presided; a man of the old-gentleman John Bull's stamp. Nothing in his face of the cast of a Jefferies. He can manage his temper, even among the vexations of law.

His Honour addressed me always with kindness. If he shampooed his summing-up, with parson's solemnity, indicating not little self-congratulation, His Honour had reason to be proud of the following remarks, which I here record for that purpose: —

"They had been told (said His Honour to the jury), that the prisoner in the dock had come sixteen thousand miles to get off from the Austrian rule — from the land of tyranny to that of liberty; and so he had, in the truest sense of the word, and that liberty which he enjoyed imposed upon him a local respect for Her Majesty, and a respect for her laws. He had the privilege of being tried by a jury, who would form their verdict solely from the facts adduced on the trial."

A fair hint; equal to saying, that under the British flag I was not going to be tried before the Holy (read, Infernal) Inquisition.

———

LXXXV.

SUNT MISERIE IN VITA HOMINIS, VIRO PROBO DOLOSIS CIRCUMDARI! NULLA MISERIA PEJOR.

MY TRIAL proceeded, before the British Jury aforesaid.

Vandemonians:

HENRY GOODENOUGH,	*Spy-Major.*
ANDREW PETERS,	*Sub-Spy.*

As an honest man, I scorn to say anything of either of you; but address myself to my God, the Lord God of Israel, in the words of Solomon: —

"If any man trespass against his neighbour, and an oath be laid upon him to cause him to swear, and the oath come before thine altar in this house: "Then hear thou in heaven, and do, and judge thy servants, condemning the wicked to bring his way upon his head." — (1 Kings viii. 31, 32.)

GEORGE WEBSTER examined: —

"I attended the meeting at Bakery-hill on the afternoon of the 29th November. Mr. Hayes was chairman, and the prisoner was on the platform. He made a speech to the effect, that he had come 16,000 miles to escape tyranny, and they (THE DIGGERS) should put down the tyrants here (POINTING TO THE CAMP). PRISONER ALSO TORE UP HIS LICENCE and threw it towards the fire recommending the others to do as he did."

N.B. — At the next state trial of Jamas Beattie, and Michael Tuhey, said witness George Webster, on his oath, was cross examined by Mr. Ireland, and stated: —

"Mr. RAFFAELLO, was at the meeting on the 29th November. — (A gold licence was here handed to the witness.) — This licence is in the name of CARBONI RAFFAELLO, and the date covers the period at which the licences were burned." — (Sensation in the Court!)

I was present in person, and a free man. AB UNO DISCE OMNES: JAM SATIS DIXI. I hereby assert that I did not burn any paper or anything at all at the monster meeting; I challenge contradiction from any *bona fide* miner, who was present at said meeting. I paid two pounds for my licence on the 15th of October, 1854, to Commissioner Amos, and I have it still in my possession. [1]

———

Examination of this gold-laced witness continued: —
"The prisoner was the most violent speaker at the meeting."
Good reader, see my speech at the monster meeting. I am sick of this witness and I will make no further comments.

1. The original Document of the following Gold-licence, as well as the Documents from DAVID BURWASH, Esq., the eminent notary-public, of 4, Castle-court, Birchin-lane, City, London; and SIGNOR CARBONI RAFFAELLO'S College Diploma, and Certificate as sworn interpreter in said City of London; together with the Originals of all other Documents, especially the letters from C. Raffaello to H. W. Archer, inserted in this book, are now in the hands of J. MACPHERSON GRANT, Esq., Solicitor, and will remain in his office, Collins-street, Melbourne, till Christmas, for inspection. — *The Printers.*

LXXXVI.

COGLIONE, IL LAZZARONE IN PARAGONE.

CHARLES HENRY HACKETT, police magistrate, cross examined by Mr. Ireland: —

"There was a deputation admitted to an interview with Mr. Rede, on Thursday night, November 30th. The prisoner was one of that deputation. I think Black was the principal party in the deputation. The deputatation as well as I remember, said, that they thought in case Mr. Rede would give an assurance that he would not go out again with the police and military to collect licences, they could undertake that no disturbance would take place. Mr. Rede replied, that as threats were held out to the effect, that in case of refusal, the bloodshed would be on their (the authorities') own heads, he could not make any such engagement at the time, nor had he the power of refraining from collecting the licence fee."

By the prisoner: —

"I recollect Commissioner Rede saying, that the word 'licences' was merely a cloak used by the diggers, and that this movement was in reality a democratic one. You (prisoner) assured him that amongst the foreigners whom you conversed with there was no democratic feeling, but merely a spirit of resistance to the licence fee."

MR. C. H. HACKETT you are a lover of truth: God bless you!
JAMES GORE, examined by the Attorney-General: —

"I am a private in the 40th, I was in the attack on the Eureka stockade. The prisoner and two other men followed me when I entered the stockade, and

101

compelled me to go out. Prisoner was armed with a pike."

Cross examined by Mr. Ireland: —

"It was day-light at the time, but not broad day-light; I had fired my musket but not used my bayonet. I ran because there were three against me. I was one of the first men in the stockade. There was no other soldier or policeman near me when the prisoner and the other men pursued me."

PATRICK SYNOTT, examined by the Attorney-General: —

"I am a private in the 40th regiment. I saw the prisoner and two other men pursuing Gore from the stockade on the morning of the attack. It was almost as lightsome at the time as it is now. I could distinguish a man at fifty yards off, and the prisoner was not fifteen yards from me. He was six or seven minutes in my sight."

JOHN CONCRITT, examined by the Attorney-General. This witness was a mounted policeman and corroborated in all particulars the evidence of the previous witnesses.
Cross examined by Mr. Ireland: —

"I fired my pistol at the prisoner. It was very good daylight. From what I saw of the soldier that morning, I should have known him again, for he stood with me for some minutes afterwards."

JOHN DONNELLY, examined by the Attorney-General: —

"I am a private of the 40th regiment. I was at the stockade on the 3rd December; I saw the prisoner there. I had a distinct opportunity of seeing."

Cross examined by Mr. Ireland: —

"I saw him for about a minute at first, and I saw him again in about ten minutes afterwards. I also saw him at the Camp the following day."

JOHN BADCOCK, trooper, examined by the Attorney-General: —

"I was at the stockade on the morning of the 3rd December. I was on foot. I snapped my musket at the prisoner, and it missed fire. I was quite close to him. I saw him again at the lock-up next day."

JOHN DOGHERTY, trooper, examined by the Attorney-General: —

"I was at the attack on the stockade. I saw the prisoner there. I knew him personally before. I have no doubt that he is the man. I saw the prisoner run towards the guard tent, and in a few minutes after, I saw him again brought back as a prisoner."

Sergeant HAGARTEY, examined by the Attorney-General: —

"I am a sergeant in the 40th. I was in the attack on the stockade. I was beside Captain Wise when he was shot. He (Captain Wise) was shot from the stockade. I saw the prisoner at the stockade. I was in the guard which took him to the Camp. The prisoner did not get away, I know. I saw him a prisoner in the Camp about five o'clock."

Cross examined by Mr. Ireland: —

"I do not know that the prisoner did not escape on his way from the stockade to the lock-up."

ROBERT TULLY, sworn and examined: —

"He was inside the stockade on the Sunday morning: saw the prisoner there armed with a pike; he was in the act of running away; saw him twice in the stockade; was sure the prisoner is the man."

Cross examined by Mr. Ireland: —

"Never saw the man before this; he was running in company with two other men; it was very early in the morning; it was some time after the stockade was taken that he was arrested; the firing then had not wholly ceased."

Private DON-SYN-GORE, drilled by sergeant HAG.
Trooper CON(S)CRIT-BAD-DOG, mobbed by Bob-tulip.
The pair of you are far below the ebb of our Neopolitan Lazzaroni!
Why did you not consult with spy Goodenough?
This having closed the case for the Crown, the Court adjourned at half-past two.

––––––––

LXXXVII.

VIRI PROBI, SPES MEA IN VOBIS; NAM FIDES NOSTRA IN DEO OPTIMO MAXIMO.

TO BE serious. I am a Catholic, born of an old Roman family, whose honour never was questioned; I hereby assert before God and man, that previous to my being under arrest at the Camp, I never had seen the face of 1, *Gore*, 2, *Synnot*, 3, *Donnelly*, 4, *Concritt*, 5, *Dogherty*, 6, *Badcock*, 7, *Hagartey*, and 8, *Tully*.
I CHALLENGE CONTRADICTION from any *bona fide* digger, who was present at

the stockade during the massacre on the morning of December 3rd, 1854.

As a man of education and therefore a member of the Republic of Letters, I hereby express the hope that the Press throughout the whole of Australia will open their columns to any *bona fide* contradiction to my solemn assertions above. I cannot possibly say anything more on such a sad subject.

LXXXVIII.

SUNT LEGES: VIS ULTIMA LEX: TUNC AUT LIBERTAS AUT SERVITUDO; MORS EMIM BENEDICTA.

ON the reassembling of the Court, at three o'clock, Mr. Ireland rose to address the Jury for the defence.

The learned Counsel spent a heap of dry yabber-yabber on the law of high-treason, to show its absurdity and how its interpretation had ever proved a vexation even to lawyers, then he tackled with some more tangible solids. The British law, the boast of *urbis et orbis terrarum*, delivered a traitor to be practised upon by a sanguinary Jack Ketch: — I., to hang the *beggar* until he be dead, dead, dead; II., then to chop the carcase in quarters; III., never mind the stench, each piece of the treacherous flesh must remain stuck up at the top of each gate of the town, there to dry in spite of occasional pecking from crows and vultures. The whole performance to impress the young generation with the fear of God and teach them to honour the king.

I soon reconciled myself to my lot, and remembering my younger days at school, I argued thus:

Where there are no bricks, there are no walls: but, walls are required to enclose the gates; therefore, in Ballaarat there are no gates. Corolarium — How the deuce can they hang up my hind-quarters on the gates of Ballaarat Township? Hence, Toorak must possess a craft which passes all understanding of Traitors.

The jury, however, appeared frightened at this powerful thundering from Mr. Ireland, who now began to turn the law towards a respectable and more congenial quarter, and proved, that if the prisoner at the bar had burnt down all the brothels not kept on the sly in Her Majesty's dominions, he would be a Traitor; yet, if he had left one single brothel standing — say, in the Sandwich Islands — for the accommodation of any of Her Majesty's well-affected subjects, then the high treason was not high — high enough and up to the mark; that is, my fore-quarter could not be *legally* stuck up on the imaginary gates of Ballaarat.

His Honour appeared to me, to assent to the line of argument of the Learned Counsel, who concluded a very lengthy but most able address, by calling on the jury to put an end by their verdict to the continued incarceration of the man, and to teach the government that they could not escape from the responsibilities they had incurred by their folly, by trying to obtain a verdict, which would brand the subjects of Her Majesty in this Colony with disloyalty.

The jury now appeared to me to be ready to let the high traitor go his way in bodily integrity.

Mr. ASPINALL then rose to address the jury on behalf of the prisoner. His speech was spirited, cutting, withering; but could only cover the falsehood, and NOT bring to light the truth: hence to record his speech here cannot possibly serve the purpose of this Book: hence the four documents, and my important observation on them in the following chapter.

LXXXIX.

MELIOR NUNC LINGUA FAVERE.

Document I.

"SUPREME COURT.
"(Before his Honour the Chief Justice,)
"The prisoner, Raffaello, on his trial being postponed, wished to address His Honour. He said that he was a native of Rome, and hoped that the same good feeling would be shown towards him in this colony as in old England. If his witnesses were there, he would be able to leave the dock at that moment. He hoped that His Honour would protect him by seeing that his witnesses were served with subpœnas.

"His Honour was not responsible for this. Prisoner's attorney was the party, and he must speak to him. It is the business of your attorney to get these witnesses."

The following advertisement appeared in *The Age,* February 24th, 1855, immediately above the leading article of said day: —

Document II.

State Trials.
"The trial of Raffaello has been postponed on account of the absence of Dr. Alfred Carr, Mr. Gordon, of the store of Gordon and M'Callum, and other witnesses for the defence. It is earnestly requested that they will be in attendance on Monday morning at latest.
"J. MACPHERSON GRANT,
"Solicitor for the defence."

The following letter, and comment on it, appeared in *The Age*, March 16th, 1855:-

Document III.

. . . "I was, Mr. Editor, present at Ballaarat on the memorable morning of the 3rd of December, and in the pursuit of my usual avocation, happened to meet Raffaello, now one of the state prisoners, on the Red-hill, he being then in search of Dr. Carr's hospital. . . . We were directed the hospital, and soon returned to the Eureka, Raffaello bringing Dr. Carr's surgical instruments. We entered the stockade, and saw many lying almost dead for want of assistance and from loss of blood, caused by gun-shot and bayonet wounds. I did not remain long in the stockade, fearing if found there at that time I would be arrested. I made my escape; but poor Raffaello, who remained rendering an act of mercy to the dying, would not leave. He might, during that time, have easily made his escape, if he wished to do so; and I am sure, ran no inconsiderable risk of being shot, through the constant explosion of fire-arms left in the stockade by the diggers in their retreat.

"J.B."

"Melbourne, 15th March, 1854."

"The writer of the above states, in a private note, that he wishes his name kept secret; but we trust that his intimacy with the Camp officials will not prevent him from coming forward to save the life of a fellow creature, when the blood-hounds of the government are yelling with anxiety to fasten their fangs upon their victims." — Ed. *A*.

The Age who certainly never got drunk yet on Toorak small-beer, had an able leading article, headed, "The State Trials" — see January 15th — concluding, "If they be found *guilty*, then Heaven help the poor State Prisoners."

Now turn the medal, and *The Age* of March 26th — always the same year, 1855 — that is, the day after my acquittal, gives copy of a Bill of the "LAST PERFORMANCE; or, the Plotters Outwitted."

Document IV.

"To-day, the familiar farce of 'STATE PROSECUTIONS; or, the Plotters Outwitted,' will be again performed, and positively for the last time; on which occasion that first-rate performer, Mr. W. F. Stawell, will (by special desire of a distinguished personage) repeat his well-known impersonation of Tartuffe, with all the speeches, the mock gravity, etc., which have given such immense satisfaction to the public on former occasions. This eminent low comedian will be ably supported by Messrs. Goodenough and Peters, so famous for their successful impersonations of gold-diggers; and it is expected that they will both appear in full diggers' costume, such as they wore on the day when they knelt before the 'Southern Cross,' and swore to protect their rights and liberties. The whole will be under the direction of that capital stage manager, Mr. R. Barry, who will take occasion to repeat his celebrated epilogue, in which he will — if the audience demand it — introduce again his finely melodramatic apostrophe to the thunder.

"With such a programme, what but an exceedingly successful farce can be anticipated? A little overdone by excessive repetition, it may be said; but still an admirable farce; and, as we have said, this is positively the last performance. Therefore, let it go on; or as Jack Falstaff says, 'play out the play.'"

Of course, I leave it to my *good reader* to guess, whether after four long months in gaol, which ruined my health for ever, I did laugh or curse on reading the above.

Concerning the four documents above, so far so good for the present; and the *Farce* will be produced on the stage of *Teatro* Argentina, Roma, by Great-works. The importance of the following observation, however, is obvious to any reader who took the proper trouble to understand the text of the first chapter of this book: —

Why Dr. A. Carr, Sub-inspector Carter, Messrs. Gordon and Binney were not present as witnesses on my trial, was, and is still, a MYSTERY to me.

Sunt tempora nostra! nam perdidi spem: melior nunc lingua favere.

XC.

PECCATOR VIDEBIT ET IRASCETUR; DENTIBUS SUIS FREMET ET TABESCET: DESIDERIUM PECCATORUM PERIBIT.

AT the end of Mr. Aspinall's able oration, the jury appeared to me, to be decidedly willing to let me go, with an admonition to sin no more: because Mr. Aspinall took the same line of defence as Mr. Michie, the counsel in the trial of John Manning; that is, he confessed to the riot, but laughed at the treason. However rashly the diggers had acted in taking up arms, however higgledy-piggledy had been the management of the stockade, yet they were justified in resisting unconstitutional force by force.

His Honour tried the patience of the jury; well knowing by experience, that twelve true-born Britons can always afford to put up with a good long yarn.

The jury retired at nine o'clock. My first thought was directed to the Lord my God and my Redeemer. Then naturally enough, to sustain my courage, I was among my dear friends at Rome and London.

To remain in the felon's dock whilst your JURY consult on your fate, is a sensation very peculiar in its kind. *To be or not to be*; that is the actual matter-of-fact question. Three letters making up the most important monosyllable in the language, which if pronounced is *life*, if omitted is *death*: an awkward position for an innocent man especially.

The jury, after twenty minutes past nine, were again in the jury-box. I was satisfied by their countenances that "the People" were victorious.

The Clerk of the Court: "Gentlemen of the Jury, have you considered your verdict?"

Foreman: "We have."

The Clerk: "Do you find the prisoner at the bar Guilty or Not Guilty?"

Foreman, with a firm voice: "NOT GUILTY!"

Magna opera Domini — (God save the People) — thus my chains sprang asunder.

The people inside telegraphed the good news to the crowd outside, and "Hurrah!" rent the air in the old British style.

XCI.

ACCIDENTI ALLE SPIE.

I WAS soon at the portal of the Supreme Court, a *free man*. I thought the people would have smothered me in their demonstrations of joy. Requesting silence, I stretched forth my right hand towards heaven, and with the earnestness of a Christian did pray as follows: — I hereby transcribe the prayer as written in pencil on paper whilst in gaol in the lower cell, No. 33.

"LORD GOD OF ISRAEL, our Father in Heaven! we acknowledge our transgressions since we came into this our adopted land. Intemperance, greediness, the pampering of many bad passions, have provoked Thee against us; yet, Oh, Lord our God, if in thy justice, Thou art called upon to chastise us, in Thy mercy, save this land of Victoria from the curse of the 'spy system.' "

Timothy Hayes answered, "Amen," and so did all the people, present, and so will my good reader answer, Amen.

XCII. & XCIII.

TO LET, No. 33, LOWER CRIBS, IN WINTLE'S HOTEL, North Melbourne.

See *Geelong Advertiser*, November 18th.
MAcKAY v. HARRISON.

Merci bien, je sors d'en prendre.

THE pair of chapters will see darkness SINE DIE; that is, if under another flag, also in another language.

GREAT-WORKS.

HESPERIA! QUANDO EGO TE AUSPICIAM? QUANDOQUE LICEBIT NUNC VETERUM LIBRIS, NUNC SOMNO ET INERTIBUS HORIS, DUCERE SOLICITÆ LICUNDA OBLIVIA VITÆ.

XCIV.

EXPLANATION,

TO BE SUBMITTED TO
HER MOST GRACIOUS MAJESTY QUEEN VICTORIA, LONDON,
AND TO
HIS HOLINESS PIUS IX., PONTIFEX MAXIMUS, ROME.
BY
MY BROTHER DON ANTONIO CARBONI, D.D.,

Head-master of the Grammar School, Coriano, Romagna.

HOMO SUM, NIL HUMANI A ME ALIENUM PUTO.

How do I explain, that I allowed one full year to pass away before publishing my story, whilst many, soon after my acquittal, heard me in person, corroborate, not indeed boastingly, the impression that I was the identical brave fellow before whose pike a British soldier was coward enough to run away.

I have one excuse, and *it is an excuse.*

The cast of mind which Providence was pleased to assign me was terribly shaken during four long, long months suffering in gaol, especially, considering the company I was in, which was my misery. The excitement during my trial, my glorious acquittal by a British jury, the hearty acclamations of joy from the people, made me put up with the ignominy and the impotent teeth-gnashing of silver and gold lace; and for the cause of the diggers to which I was sincerely attached, *I was not sorry* at the Toorak spiders having lent me the wings of an hero — the principal foreign hero of the Eureka stockade. My credit consists now in having the moral courage to assert the truth among living witnesses.

"And I proposed in my mind to seek and search out wisely concerning all things that are done under the sun. This painful occupation hath God given to the children of men to be exercised therein. I have seen all things that are done under the sun, and behold all is vanity and vexation of spirit." — *The Preacher,* chap. 1st, v. 13, 14.

XCV.

QUI POTEST CAPERE CAPIAT.

ELECTION.
OLD SPOT, BAKERY-HILL, BALLAARAT.

ACCORDING to notice, a Public Meeting was held on Saturday, July 14th, 1855, for the election of nine fit and proper men to act as Members of the Local Court — the offspring of the Eureka Stockade.
The Resident Warden in the Chair. Names of the Members elected for the FIRST LOCAL COURT, Ballaarat: —

I. JAMES RYCE	
II. ROBERT DONALD	
III. CARBONI RAFFAELLO	ELECTED UNANIMOUSLY
IV. JOHN YATES	
V. WILLIAM GREEN	
VI. EDWARD MILLIGAN,	elected by a majority of 287 votes.
VII. JOHN WALL,	elected by a majority of 240 votes.
VIII. THOMAS CHIDLOW,	elected by a majority of 187 votes.
IX. H. R. NICHOLLS,	elected by a majority of 163 votes.

The first time I went to our Court, I naturally stopped under the gum-tree — before the Local Court Building — at the identical spot where Father P. Smyth, George Black, and myself delivered to the Camp authorities our message of peace, for preventing bloodshed, on the night of Thursday, November 30th, 1854, by moonlight. We were *then* not successful.

Now, I made a covenant with the Lord God of Israel, that if I comparatively regained my former health and good spirits, I would speak out the truth; and further, during my six months' sitting in the Court, I would give right to whom right was due, and smother the knaves, irrespective of nationality, religion, or colour.

I kept my word — that is, my bond is now at an end.

I hereby resign into the hands of my fellow-diggers the trust reposed in me as one of their arbitrators: after Christmas, 1855, I shall not sit in the Local Court. With clean hands I came in, with clean hands I go out: that is the testimony of my conscience. I look for no other reward.

(Signed)   CARBONI RAFFAELLO.

Dec. 1st, 1855.

XCVI.

EST MODUS IN REBUS: SUNT CERTI DENIQUE FINES, QUOS ULTRÆ, CITRAQUE NEQUIT CONSISTERE RECTUM.

HAVE I anything more to say? Oh! yes, mate; a string of the realities of the things of this world.

Some one who had been spouting, stumping, and blathering — known as moral-force "starring" — in *urbe et argo,* for the benefit of the state prisoners, had for myself personally *not* humanity enough to attend to a simple request. He could afford to ride "on coachey," I had to tramp my way to Ballaarat. I wished him to call at my tent on the Eureka, and see that my stretcher was ready for my weary limbs.

Full stop. My right hand shakes like a reed in a storm; my eyes swell from a flood of tears. I can control the bitterness of my heart, and say, "So far shalt thou go;" but I cannot control its ebb and flow: just now is spring-tide.

If I must again name a noble-hearteded German, Carl Wiesenhavern, of the Prince Albert Hotel, who was my good Samaritan, I must also annex the following three documents, because my friends in Rome and Turin may take my wrongs too much to heart!

XCVII.

THE END OF MEN WHOSE WORD IS THEIR BOND.

(Per favour of *The Times.*)

"ON the disgraced Sunday morning, December 3rd, whilst attending the wounded diggers at the London Hotel, I was arrested by seven troopers, handcuffed, and dragged to the Camp. On my arrival there, I was commanded to strip to the bare shirt; whilst so doing I was kicked, knocked about, and at last thrown into the lock-up by half-drunken troopers and soldiers. My money, clothes, and watertight boots, which were quite new, could nowhere be found at the Camp. Gaoler Nixon had bolted.

"From the confusion and excitement of that morning, I cannot say with certainty the whole extent of my loss; but I can conscientiously declare that it amounted to £30. The only thing which I saved was a little bag, containing some Eureka dust, and my 'Gold-licence,' which Inspector Foster, who knew me, took charge of previous to my ill-treatment, and has subsequently handed over to Father P. Smyth for me.

"Awaiting my trial in the Melbourne gaol, I made my 'complaint' to the visiting justice, for the recovery of my property; but as I had not even a dog to visit me in prison, so my complaint remained unnoticed. After all, said worshipful, the visiting justice (who was ushered into our yard with 'Fall in, hats off!'), needs more power to him, as *Joseph*, the nigger-rebel, for the £8, which had been robbed from him in due form at the Camp, had the consolation to be informed by his worshipful that gaoler Nixon had bolted.

"The glorious 'Not Guilty' from a British jury having restored me to my former position in society, I embodied my 'claim' for restitution in a constitutional form, and had it presented by a gentleman to the Colonial Secretary, to be submitted for his Excellency's KIND Consideration. His Excellency, soon after my trial, on being assured of my testimonials to character and education, condescended to say, 'He was glad to hear I was so respectable;' but His Excellency has not yet been pleased to command the restitution of my property.

"Disappointed, in bad health, and worse spirits, I tramped for Ballaarat, where I found that my tent, on the Eureka, had been robbed of everything that was worth literally a sixpence — cradle, two tubs, digging tools, cooking utensils, all gone, even my very blankets! and, of course, all my little gold in specimens and dust, as well as my belt with money in it.

"From my account-book I can positively say, that on the fatal morning I was arrested, the money I had on my person, and what I had in my tent in real cash, was £49. ALL OF WHICH I had earned by the sweat of my brow, honestly, through downright hard work.

"During the whole of last season, on the Eureka, who was the first every morning, between four and five to sing out 'Great works?' Who was the last dilly-dallying at the cradle after sunset? I appeal to my fellow-diggers, and with confidence.

"Brooding over the strange ups and down of life, I found some consolation in the hearty cheers with which I was saluted at the Adelphi Theatre for my song —
'When Ballaarat unfurled the Southern Cross;'
and I had the peculiar sensation on that particular night to lie down on my stretcher very hungry!

"'Heu mihi! pingui quam macer est mihi taurus in arvo!' and it must be acknowledged that it would have been paying an honest and educated man a better compliment if my neighbours on the Eureka had found less edification in witnessing my nice snug tent converted into a gambling house by day, and a brothel by night. A sad reflection! however merry some scoundrels may have made in getting drunk with my private brandy in the tent.

"Never mind! the diggers have now a legion of friends. So I prevailed on myself to tell, half-a-dozen times over to most of the 'well-disposed and independent' yabber-yabber leaders on Ballaarat, how I had been robbed at the Camp, how for my sorrows every mortal thing had been stolen from my tent, and concluded with the remark, 'that in each case the thieves were neither Vandemonians nor Chinese.'

"I met with grand sympathy in 'words,' superlatively impotent even to move for the restitution of my watertight boots!

"Hurrah! glorious things will be told of thee, Victoria!

"These waterhole skippers, who afford buzzing and bamboozling when the rainbow dazzles their dull eyes, bask in their 'well-affected' brains the flaring presumption that 'shortly' there will be a demand for sheeps' heads! (Great works!) and pointing at several of us, it is given unto them to behold with glory 'the end of men whose word is their bond!' (Great works!)

"Let us sing with Horace —

'TUNE — Old Style.

'Quando prosperus et jucundus,
Amicorum es fecundus,
Si fortuna perit,
Nullus amicus erit.
    Chorus — Cives! Cives!
        Querenda pecunia primum,
        Post nummos virtus.

"Which in English may mean this —

'A friend in need is a friend indeed,' that's true,
    But love now-a-days is left on the shelf,
The best of friends, by G—— in serving you
    Takes precious care first to help himself.
Ancestors, learning, talent, what we call
Virtue, religion — MONEY beats them all.

"I must now try the power of my old quill, perhaps it has not lost the spell—

"In Rome, by my position in society, and thorough knowledge of the English language, I was now and then of service to Englishmen THERE; in my adversity is there a generous-hearted Englishman HERE who would give me the hand and see that the government enjoins the restitution of the property I was robbed of at the Camp. Let the restitution come from a Board of Inquiry, a Poor-law Board, a Court-Martial, or any Board except a Board (full) of Petitions. The eternal petitioning looks so 'Italian' to me! And, especially, let the restitution of my new water-tight boots be done this winter!

"As for the ignominy I was subjected to, my immense sufferings during four long, long months in gaol, the prospects of my life smothered for a while, we had better leave that alone for the present.

"Were I owned by the stars and stripes, I should not require assistance, of course not; unhappily for the sins of my parents, I was born under the keys which verily open the gates of heaven and hell; but Great Britain changed the padlocks long ago! hence the dreaded 'Civis Romanus sum' has dwindled into 'bottomed on mullock.'

"CARBONI RAFFAELLO,
"By the grace of spy Goodenough Captain of Foreign Anarchist.
"Prince Albert Hotel, Ballaarat,
"Corpus Christi, 1855."

No one did condescend to notice the above letter. I do not wonder at it, and why?
I read in the Saturday's issue of *The Star*, Ballaarat, October 6th, 1855, how a well-known digger and now a J.P., did, in a " Ballaarat smasher," toast the good exit of a successful money-maker — an active, wide-awake man of business certainly, but nothing else to the diggers of Ballaarat: — *"Cela n'est pas tout-a-fait comme chez nous."*

XCVIII.

SUNT TEMPORA NOSTRA!

THAT IS THE FOLLOWING FROM TOORAK.

"Colonial Secretary's Office, Melbourne,
"October 8th, 1855.
"Sir, — Adverting to your correspondence (September 30th), in reply to my letter of the 20th ultimo, I am directed by His Excellency to state that government are compelled to adhere to fixed rules — THEY BY NO MEANS DOUBT THE VERACITY OF YOUR STATEMENT, but they have a duty to the public to perform, which imposes the necessity of never granting money in *compensation*, except when the clearest evidence of the loss is given, and that a personal statement no matter by whom given, is never accepted as sufficient testimony.

"I have the honour to be,
"SIR,
"J. MOORE, A. C. S."

"MR. CARBONI RAFFAELLO,
"Gravel-pits, Ballaarat-flat.

A *Cheer-up* written for the MAGPIE of BALLAARAT, perched on the Southern Cross Hotel, Magpie-gully.

No more from MOORE;
Too dear! his store.
Hang the "Compensation:"

Speak of "RESTITUTION!"
Do not steal! 's an old Institution,
Restiuere?　Popish innovation.

CHORUS.

COO-HEE!　Great works at Toorak!
COO-HEE!　Keep clear of th' WOOL-pack.

WATERLOOBOLTER CHIMES.

SIP	sop	stir-up	Toorak	small	beer
do	*si*	*la sol*	*fa me*	*re*	*do*
Nip	nap	wash down	chops nacks	oh!	dear

———

XCIX.

SUPPOSE I give now the kind (!) answer from Police-inspector HENRY FOSTER! it
will give general satisfaction, I think: —

"Police Department,
　" Ballaarat, Nov. 2, 1854.
"Sir, — In reply to your communication, dated 26th ultimo, on the subject
of your having been deprived of your clothing during your arrest at this
Camp, in December, 1855 [*I think, Mr. Foster, it was in 1854*] I have the honour
to inform you that, to the best of my recollection, the clothing you wore when
you were brought to the Camp consisted of a wide-awake hat, or cap, a red
shirt, corduroy or moleskin trowsers, and a pair of boots.

"Of these articles, the cap, shirt, and boots were put amongst the surplus
clothing taken from the other prisoners, and I am not aware how they were
disposed of afterwards.

"I must add, that the shirt alluded to was made of wool, under which you
wore a cotton one, the latter of which you retained during your confinement.
　　　　"I have the honour to be, Sir,
　　　　　　"Your obedient servant,
　　　　　　　　"HENRY FOSTER,
　　　　　　　　　"Inspector of Police.
"SIGNOR CARBONI RAFFAELLO."
" Ballaarat.

115

My money is not mentioned though! Very clever: and yet I know it was not Foster who did rob me.

However, good reader, if you *believe* that a Ballaarat miner, of sober habits and hard at work, has not got about his person, say a couple of £1 *rags*, well . . . there let's shut up the book at once, and here is the

END.

P.S. If John Bull, cross-breed or pure blood, had been *robbed* in Italy, half less wantonly, and twice less cruelly, than myself, the whole British press and palaver *in urbe or orbe terrarum* would have rung the chimes against Popish *gendarmes* and the holy (!) inquisition of the scarlet city. So far so good.

A friendless Italian is ROBBED under arrest on British ground, close by the British flag, by British troopers and traps: oh! that alters the case.

What business have these foreign beggars to come and dig for gold on British Crown lands?

BASTA COSI; *that is*, Great works!

C.

WANTED — STUFF, ANYHOW, FOR THE LAST CHAPTER.

IF *The Age*, always foremost in the cause of the digger, never mind his language or colour; if *The Argus* would drop the appending "a foreigner" to my name, and extend even unto me the old motto "fair-play;" if *The Herald* would set up the pedestal for me whom it has erected as a "MONUMENT OF GRATITUDE;" I say, if the gentlemen Editors of the Melbourne Press, on the score of my being an old *Collaborateur* of the European Press, would for once give a pull, a strong pull, and a pull altogether, to drag out of the Toorak small-beer jug, the correspondence on the above matter between

1. SIR CHARLES HOTHAM, K.C.B.
2. W. C. HAINES, C.S.
3. W. FOSTER STAWELL, A.G.
4. MR. STURT, Police Magistrate.
5. W. H. ARCHER, A.R.G.
6. CAPTAIN M'MAHON.
7. POLICE-INSPECTOR H. FOSTER.
8. Another whom I detest to name, and
9. SIGNOR CARBONI RAFFAELLO, M.L.C. of Ballaarat,

it would astonish the natives, teach what emigration is, and I believe the colony at large would be benefitted by it.

There are scores of cases similar to mine, and more important by far, because widows and orphans are concerned in them. *Sunt tempora nostra!*

Master *Punch*, do join the chorus; spirited little dear! won't you give a lift to Great-works? Spare not, young chip, or else, the jackasses in the Australian bush will breed as numerous as the locusts in the African desert.

It is not FEAR that makes me shake at chapters XCII and XCIII. Good reader, to the last line of this book, my quill shall stick to my word as given in the first chapter. Hence, for the present, this is the LAST. Put by carefully the pipe, we may want it again: meanwhile, FAREWELL.

———————

www.ingramcontent.com/pod-product-compliance
Lightning Source LLC
Chambersburg PA
CBHW051732040426
42447CB00008B/1101